WHEN WAS LATIN AMERICA MODERN?

Studies of the Americas

Edited by
James Dunkerley
Institute for the Study of the Americas
University of London
School of Advanced Study

Titles in this series are multidisciplinary studies of aspects of the societies of the hemisphere, particularly in the areas of politics, economics, history, anthropology, sociology and the environment. The series covers a comparative perspective across the Americas, including Canada and the Caribbean as well as the USA and Latin America.

Titles in this series published by Palgrave Macmillan:

Cuba's Military 1990–2005: Revolutionary Soldiers during Counter-Revolutionary Times
By Hal Klepak

The Judicialization of Politics in Latin America
Edited by Rachel Sieder, Line Schjolden, and Alan Angell

Latin America: A New Interpretation
By Laurence Whitehead

Appropriation as Practice: Art and Identity in Argentina
By Arnd Schneider

America and Enlightenment Constitutionalism
Edited by Gary L. McDowell and Johnathan O'Neill

Vargas and Brazil: New Perspectives
Edited by Jens R. Hentschke

When Was Latin America Modern?
Edited by Nicola Miller and Stephen Hart

Debating Cuban Exceptionalism
Edited by Laurence Whitehead and Bert Hoffman

Caribbean Land and Development Revisited
Edited by Jean Besson and Janet Momsen

The Hispanic World and American Intellectual Life
By Iván Jaksic

Bolivia: Revolution and the Power of History in the Present
By James Dunkerley

Brazil, Portugal, and the Black Atlantic
Edited by Nancy Naro, Roger Sansi-Roca and David Treece

The Role of Mexico's Plural in Latin American Literary and Political Culture
By John King

Democratization, Development, and Legality: Chile 1831–1973
By Julio Faundez

The Republican Party and Immigration Politics in the 1990s: Take Back Your Tired, Your Poor
By Andrew Wroe

Faith and Impiety in Revolutionary Mexico
By Matthew Butler

When Was Latin America Modern?

Edited by
Nicola Miller and Stephen Hart

WHEN WAS LATIN AMERICA MODERN?
© Nicola Miller and Stephen Hart, 2007.
Softcover reprint of the hardcover 1st edition 2007 978-1-4039-8000-7

First published in 2007 by
PALGRAVE MACMILLAN™
175 Fifth Avenue, New York, N.Y. 10010 and
Houndmills, Basingstoke, Hampshire, England RG21 6XS
Companies and representatives throughout the world.

PALGRAVE MACMILLAN is the global academic imprint of the Palgrave Macmillan division of St. Martin's Press, LLC and of Palgrave Macmillan Ltd. Macmillan® is a registered trademark in the United States, United Kingdom and other countries. Palgrave is a registered trademark in the European Union and other countries.

ISBN 978-1-349-53864-5   ISBN 978-0-230-60304-2 (eBook)
DOI 10.1057/9780230603042

Library of Congress Cataloging-in-Publication Data

When was Latin America modern? / edited by Nicola Miller and Stephen Hart.
p. cm.
Includes bibliographical references and index.

1. Latin America—Civilization—Congresses. 2. Civilization, Modern—Congresses. I. Miller, Nicola. II. Hart, Stephen M.

E1408.3.W46 2007
980—dc22   2006050734

A catalogue record for this book is available from the British Library.

Design by Newgen Imaging Systems (P) Ltd., Chennai, India.

First edition: April 2007

10 9 8 7 6 5 4 3 2 1

Transferred to Digital Printing 2011

Contents

Introduction: Interdisciplinary Approaches to Modernity
in Latin America   1
Nicola Miller

Part I Views from the Historical and Social Sciences

1 Geographies of Modernity in Latin America:
  Uneven and Contested Development   21
  Sarah A. Radcliffe

2 Modernity and Tradition: Shifting Boundaries,
  Shifting Contexts   49
  Peter Wade

3 Mid-Nineteenth-Century Modernities
  in the Hispanic World   69
  Guy Thomson

4 When Was Latin America Modern? A Historian's
  Response   91
  Alan Knight

Part II Views from Literary and Cultural Studies

5 When Was Peru Modern? On Declarations
  of Modernity in Peru   121
  William Rowe

6 Belatedness as Critical Project: Machado de Assis
  and the Author as Plagiarist   147
  João Cezar de Castro Rocha

7 Cuban Cinema: A Long Journey toward
  the Light   167
  Julio García Espinosa

CONTENTS

8 Culture and Communication in Inter-American
   Relations: The Current State of an
   Asymmetric Debate   177
   Néstor García Canclini

Conclusion: When Was Latin America Modern?   191
Laurence Whitehead

List of Contributors   207

Index   211

Introduction

Interdisciplinary Approaches to Modernity in Latin America

Nicola Miller

The question was designed to provoke, and indeed it did. Debates at the interdisciplinary event behind this book turned not so much on any putative answers but on the validity of the question itself, the criteria for addressing it, and whether the concept of modernity could be given any meaningful analytical content at all. The strongest case against the term was made by anthropologist Peter Wade, for whom modernity's connotations of teleology (the inevitability of the grand march of progress) and scaling (modernity as the big picture, with everything non-modern diminished to the small and insignificant) were too inescapable to make it useful or acceptable as a tool of analysis, even in the variant of "multiple modernities." Most of the participants, however, ultimately declared in favor of retaining the term and debating strategies for endowing it with analytical substance, although the range of referents in this collection of essays (from Enlightenment to ephemerality) is in itself testimony to the problems involved in trying to do so.

Latin America's experience of modernity has been the subject of much academic attention over the past two decades, both from Latin American scholars and from other Latin Americanists. Interest has arisen at least partly in response to debates about the relevance of postmodernism and the impact of that constellation of changes customarily referred to as "globalization" (e.g., Quijano 1990; Rincón 1995; Sáenz 2002, among others). The question of modernity has been especially prominent in the field of cultural studies, but many historians, social scientists, and specialists in film, the built environment, and the visual arts have also organized their work around the theme of what it meant to be modern in Latin America. The resulting literature on the topic is rich, but it is also—as Latin America's modernity itself is often claimed to be—fragmented. Scholars from different disciplines (and, indeed, within each discipline) have taken widely varying positions on fundamental issues such as the chronology of modernity, its character and its agents. It would be impossible to review the whole of this literature here, but a rapid sampling of key works from Latin Americans who have worked on "modernity" over the last decade or so should convey a sense of the variety of views in play. When was Latin America modern? If we take the question literally, the case has been made for the late fifteenth century, with the onset of European imperialism (mainly by philosophers, e.g., Quijano 1990; and Dussel 1995); the early nineteenth century, with the independence struggles (mostly by historians, such as Guerra 1992, 1995; and Uribe Urán 2001); the late nineteenth century, with integration into the international economy and the emergence of *modernismo* (the focus of literary and cultural studies: see, e.g., Rama 1984; Ramos 1989; Jrade 1998; Geist and Monleón 1999); the mid-twentieth century, with the spread of mass technologies (social scientists such as Brunner, 1995); and the late twentieth century, with neoliberalism and democracy (political scientists and economists too numerous to list), not to mention several other periods in between.

To an extent, of course, the divergence of views about Latin America's modernity is only one manifestation of the lack of consensus about when a consciousness of modernity in general emerged: many scholars, particularly historians, gravitate toward the late eighteenth and early nineteenth centuries (the impact of Enlightenment models of rationality coupled with the ascendancy of capitalist practices; see, e.g., Geras and Wokler 2000; Bayly 2003), but others have made valid arguments about the seventeenth century (the Reformation and the scientific revolution), and as early as 1200 has been proposed (Hardt and Negri 2000). However, such widespread uncertainties about the history of modernity still have less immediate political significance in countries widely regarded as already being modern (even if this is a condition that needs constant vigilance to maintain it), than in countries that are not regarded thus. Renato Rosaldo has commented upon the "absolute ideological divide" between so-called modernized and non-modernized nation-states, noting that in the United States issues such as high infant mortality rates among African Americans "are treated neither as signs of underdevelopment nor as failures of uneven modernization (as they conceivably could be in principle and no doubt would be in Latin America)" ("Foreword" to García Canclini 1995: xiii–iv). The oft-heard propositions that Latin America has had modernism without modernization, or modernization without modernity, or an experience of modernity that was no more than a pseudo-modernity, are all based on what Mark Thurner has called "the metanarrative of the deficient," that is, the persistent sense that Latin America's distinctive history of early decolonization and early experience of neocolonialism is somehow lacking because it does not correspond to patterns discerned elsewhere. Partly to counter such assumptions, there have also been several variations on the theme that Latin America has developed an alternative modernity, which, it is increasingly claimed, has sustained the original emancipatory impulse of the independence wars and offers a model for contending with twenty-first-century change.

It was in order to debate the variety of views outlined here and the theoretical and methodological issues surrounding them that we convened an interdisciplinary workshop, which was held in February 2005 at the Institute for the Study of the Americas in London. Bringing together scholars from different disciplines to discuss a topic of common interest is both fashionable and hazardous. Interdisciplinary events are looked upon favorably by funding bodies and attract a lot of interest in the academic community: there is a widespread—albeit vague—sense that they are a good thing. But they do not, as Barthes put it, offer "the calm of an easy security" (1977, 155), at least not if they actually achieve their objective of stimulating people to engage with the contributions of other disciplines (and many academics are all too familiar with attending events where the historians comment only on the history papers, the anthropologists only on the anthropological papers, and so forth, and each discipline sails serenely on its way, wholly unperturbed by the shock waves of other epistemologies, but given a fair wind by the satisfaction of having been dutifully "interdisciplinary"). Even when engagement does take place, as it did on this occasion, connections between disciplines can prove elusive, and it is often the differences not only between disciplines but also within them that come to the fore. Or it transpires that some connections are made but not others: historians are often willing to learn from political scientists or geographers, but find it harder to appreciate the relevance to them of cultural studies; anthropologists and cultural studies people tend to find a lot in common (partly because many of them are drawing on a common body of theoretical literature; and at least some of them share a basically ethnographic methodology), but both find it difficult to contend with the residual positivism of even the most theoretically aware historians.

To a greater or lesser extent, all of the aforementioned tensions were evident at the "When Was Latin America Modern?" workshop.

Thus, the following collection of essays has a strong thematic integrity, but also illustrates the dramatic variety of approaches to the question of modernity. There are also absences (unavoidable for logistical reasons): we had no papers by philosophers, art historians, or economists, to mention only the most obvious of the other disciplines that could have been involved. Even so, the editors of this volume maintain that the exchanges from that encounter ultimately succeeded in doing more than providing the always-valuable incentive for disciplines to reflect critically upon themselves. In what follows, I identify some of the convergences that could create a basis for rethinking approaches to Latin American modernity in the light of interdisciplinarity. First, though, I outline the chapters, which are presented in two parts: I. Views from the Historical and Social Sciences (chapters 1, 2, 3, and 4); and II. Views from Literary and Cultural Studies (chapters 5, 6, 7, and 8).

Outline of the Chapters

Chapter 1 by Sarah Radcliffe, "Geographies of Modernity in Latin America: Uneven and Contested Development," shows how rich a perspective the discipline of geography, informed by social theory and cultural studies, can offer. Keenly sensitive to the normative implications of the term modernity, she begins with an overview of geographical approaches to the topic, drawing out a critique of the existing literature in order to develop a new framework for understanding Latin American modernity. She goes on to illustrate this with case-study evidence from Ecuador. Starting from the premise that modernity is a concept with inescapably geographical connotations, she argues that even so it does not necessarily entail either Eurocentrism or diffusionism. Adopting Pred's and Watt's concept of "multiple reworked modernities" (1992), she emphasizes that acknowledging that European versions of modernity have historically been the dominant form does not necessarily mean that they have to be regarded as a universal standard. Breaking down the monolith of modernity to establish an analytical framework of project, discourse, and experience, she argues that the key question is not "when was Latin America modern?," which will deterministically confine the inquiry to a teleological, normative approach, in which Latin America can only be seen as a late arrival at modernity's ball, but rather "in which spaces was Latin America modern?" She invokes the metaphor of the fulcrum to capture the shifts in interconnectedness and differentiation, solidarity and hierarchy, all of which are components of the constructions of modernities in specific times and spaces. Neither of the two main forces conventionally seen as driving modernity—capitalism and the nation-state—are monolithic, she emphasizes. There is no predictable or regular link between modernity and development: reverses can occur, as Latin America's experiences during the 1980s showed all too acutely. Radcliffe's work makes a compelling case for the significance of geographies—both territorial and imaginative—in the construction of modernity, and for taking into account the bodily aspect of the experience of modernity as well as the mental processes that so often absorb the attention of historians and cultural critics.

For Peter Wade, Latin America has always—or never—been as modern as anywhere else. His radical doubts about the analytical value of the term modernity stem from his concerns about the persistence of dualistic ways of thinking in Western social science, particularly within his own discipline of anthropology. For him, as discussed in chapter 3, "Modernity and Tradition: Shifting Boundaries, Shifting Contexts," the workshop's question entailed the assumption of a historical narrative dominated by a Western modernity that blazed the trail of progress and prosperity, leaving all other societies limping along haltingly in its wake, and determining the context in which Latin America defines itself and is defined by others, both temporally and spatially. He discusses two commentaries on García Canclini's *Hybrid Cultures* (1992) in order to illustrate how difficult it is to eliminate an underlying teleology and a scaling effect in which modernity is writ large scale and global while tradition is rendered small scale and local. Even though he takes the view, contrary to some critics, that García Canclini's text does destabilize both "tradition" and "modernity," he notes that the very possibility of the hybrid implies that we know what was traditional and what was modern in the first place (just as *mestizaje* is dependent upon a notion of racial purity, as Wade has argued elsewhere). More optimistically, he argues that anthropology can also offer ways to undo these dualisms and their underlying Eurocentrism. Drawing on the biological concept of natural selection, he suggests that a model of complex networks, involving nonlinearity and internally generated organization, can act as a source of inspiration for thinking in non-scalar and non-teleological ways, even though social scientists would have to find some way of accounting for human agency. He goes on to offer a series of examples of how approaches placing more emphasis on multilateralism and mutual exchange might work in particular analytical contexts. In his own research, particularly on music in Colombia, he found that the supposedly traditional—a category to which great significance was attached by all those involved—was often as hybridized and as modern as the modernity with which it was forming hybrids. Overall, Wade's radical skepticism about the analytical value of the term modernity acts as a compelling reminder that even though academics now make ritual obeisance to the idea that our categories of analysis, such as global and local, modernity and tradition, are ways of reading and construing processes of change rather than objective realities in themselves, in practice it is not always easy to keep this in sight.

Historian Guy Thomson starts chapter 4, "Mid-Nineteenth-Century Modernities in the Hispanic World," by discussing a historical example of precisely what Wade was talking about: the construction of categories by anthropologists to suit their own ends. In rural Mexico during the 1920s, U.S. anthropologists built models of cultural change based on a conception of modernization as inevitable, but in the process they gathered much empirical evidence about the presence of "modern" practices and goods. Drawing not on their analytical model but on the kind of evidence about subjective and cultural experience that lay behind it, Thomson adopts a similarly local-level perspective to compare two regions in Mexico and Spain where, he argues, a consciously experienced modernity was felt in the mid-nineteenth century. Thomson goes on to discuss the potential strengths and weaknesses of comparative history as a methodological approach, taking as a case in point C. A. Bayly's *The Birth of the Modern World 1780–1914* (2003). Bayly's premise was that being modern was at least partly a process of self-definition; therefore, evidence about subjective experience had to be taken into account—an approach that Thomson found inspiring. Yet supposedly "global" histories are necessarily selective. The Hispanic world is largely omitted from Bayly's book, and when it is mentioned it is invariably in stereotypical terms that are bound to strike any Latin Americanist as astounding, especially given the author's sensitivity to comparable situations in Asia. In this respect, Bayly's approach illustrates the force of Wade's points about the dangers of retaining perspectives of teleology and scaling, even when it is resistance to the dominant model that is being privileged in the discussion. Thomson then goes on to demonstrate, in his own carefully documented reconstruction of everyday life in Puebla Sierra and the Málaga-Granada highlands from the 1850s to the 1870s, how comparative history can give very precise content to a concept of modernity. For Thomson, whose approach integrates economic, political, social, and cultural factors, modernity entails a culture of consumption, secular associational life, and the politicization of traditional solidarities around democratic ideals. His evidence about the presence of civic associations and democratic practices in public everyday life is supported by Carlos Forment for Mexico and Peru, and prospectively for Argentina and Cuba too (Forment 2003). Thomson's chapter, which has the advantage of comparing a Latin American and a non-Latin American example, provides ample evidence of the limitations of any teleological model—movements toward modernity can go back as well as forward, as was shown by events in Mexico after the restoration of the Republic in 1867, when the previously increasing belief in democracy and economic progress was tempered by authoritarian reaction. A culture of consumption had been created without a corresponding culture of citizenship. Thomson's case studies illustrate the possibilities of thinking non-teleologically, but whereas Wade's logic leads him to jettison the concept of modernity altogether, Thomson strives to give it specific content within a comparative framework.

Chapter 5 by Alan Knight, "When Was Latin America Modern?: A Historian's Response," directly confronts the conceptual difficulties attendant on the question "When was Latin America Modern?" and comes to a skeptical conclusion as to its validity as a heuristic device (although granting it an instrumental value in stimulating debate). The focus of his concerns is different from Wade's, however. From his point of view, careful attention to historical process can counter the dangers of teleology and scaling (which are inherent in many of the concepts historians habitually use); the real problem with the term is that it is very difficult to give it any meaning that is both consistently applicable and rigorous. He draws attention to the valuable distinction (drawn from linguistics via anthropology) between the "emic" and the "etic," or how concepts are understood by actors in a specific historical context and how they are used by social scientists in their analyses of those actors. He argues that the term "modernity" was not used in Latin America until the late twentieth century (and then primarily in academic discourse). In addressing the issue of the specific analytical content of the terms "modern" and "modernity," however, he challenges those who see modernity in Latin America as primarily alien and imported, arguing that any such model both "neglects multiple invention and discovery" and denies Latin America any "autonomous capacity to generate its own modernity" (p. 98). If modernity means anything, he suggests, it refers to the package of ideas and assumptions known as the European Enlightenment. Even then, the situation is far from clear, for in many parts of Europe itself, let alone in other parts of the world, the history of the spread of those ideas is "one of selective appropriation, distortion, and repudiation"

(p. 101). It is possible, he suggests, and possibly even useful, to trace manifestations of these ideas in various regions of various countries of Latin America, at various times. To go further, however, to try to determine when Latin America became modern, is, he concludes, to apply an ill-defined concept to a necessarily limited set of data.

Chapter 6 by William Rowe, "When Was Peru Modern? On Declarations of Modernity in Peru," displays the insights that can be gleaned from taking up the challenge to escape the confines of linear thinking. The narrative of progress has been so unquestioned an assumption of post-Enlightenment intellectual models, he suggests, that "to think temporal heterogeneity requires an act of will" (p. 140). As José Carlos Mariátegui famously argued, the Eurocentric Marxist framework of history in stages, progressing from feudalism to capitalism to socialism, could not meaningfully be applied to Peruvian realities: the only way to overcome the view of Peru's history as lacking was to redeem the past by a willed projection of it into the future. Identifying a correspondence between Mariátegui's ideas about history and Walter Benjamin's discussion of the possibilities for reading the relationship between the past and the present, Rowe develops a critique of conventional historical method, with its adherence to sequential narrative. To do so, he explores various scenes, from nineteenth- and twentieth-century Peruvian literary and historical texts, in which various recognizably modern senses of temporality are constituted. His idea is that the sections of his chapter, each of which evokes one particular scene of modernity, can be read in any order: they are conceived as a constellation, not as a continuum. The presentation of the material thus—negating seriality and sequentiality; bringing together temporality and spatiality—both enacts and illustrates his main argument that a nonlinear approach is necessary to understanding the idea of modernity in Peru. Historicity matters, but so does cultural distinctiveness. Images of modernity are generated not only at a particular time but also in a specific space. It may be, he implies, that the only way to approach modernity in Latin America is to retain a keen sense of awareness that there will always be a gap, an insufficiency in what can be known. Peru has been "simultaneously modern and non-modern," (p. 130) and anyone who analyzes the country's history needs to find a heuristic device sufficiently flexible and self-critical to encompass that multiple reality.

Instead of focusing directly on the value of modernity as an analytical term, João Cezar de Castro Rocha approaches the problem from a different angle, discussing it in what Alan Knight calls "emic" terms, that is, how it was understood in the specific historical context of late nineteenth-century Brazil. Then and there, one key component of modernity, along with economic progress and social justice, was the desire to be up-to-date with the latest trends in the central powers. Just because elements of teleology and scaling were thereby embedded in the emic experience, however, does not mean that they are necessarily carried over into the "etic" analysis, as chapter 7 by de Castro Rocha, "Belatedness as Critical Project: Machado de Assis and the Author as Plagiarist," shows. In discussing how the major Brazilian novelist Machado de Assis responded to modernization, de Castro Rocha also offers an analytical approach to modernity that helps to go beyond the fact that teleology and scaling tend to be built in to projects of modernity themselves, thereby making it even harder for the would-be analyst of modernity to shed those assumptions. What his case study shows is that even what might appear to be highly constrained circumstances—in this instance, the effects of traumatic modernization in the context of what has always been read as the oppressive historical bind of civilization versus barbarism—can actually produce radically different outcomes. While not denying that a fatalistic response engendering pessimism and/or repression was and has remained a major factor in Latin American cultural politics, de Castro Rocha illustrates how Machado de Assis developed a response that was optimistic and creative. Rather than seeing the impossibility of originality as disempowering, Machado explored the idea that it was in practice the opposite because it liberated him from relating himself to any particular tradition and opened up the possibility of appropriating any and every tradition. In other words, he accepted his location as always already behindhand, but reinterpreted it as an advantage not a hindrance. Thus, by adopting a strategy that de Castro Rocha calls "belatedness as critical project," Machado becomes a conscious plagiarist, undermining existing (especially Romantic) concepts of authorship, drawing attention to the extent to which all writers, everywhere, are first of all readers, and demonstrating the inadequacy of analytical frameworks of Latin American culture based on the "anxiety of influence."

Julio García Espinosa, in chapter 8 "Cuban Cinema: A Long Journey Towards the Light," brings a cinematographer's eye to the issues, looking at modernity through the frame of Cuban film, in which, as is well known, he himself has played a highly significant role as a pioneering director and a joint founder (with Tomás Gutiérrez Alea) of the Cuban Film Institute (ICAIC). His account of the rise and development of Cuban cinema from virtual nonexistence before the Revolution to playing a key role in making not just Cuba but also

Latin America as a region *visible* is yet another telling instance of how close historical analysis can illuminate the importance of a convergence of conditions in bringing about modernity. He and Gutiérrez Alea, returning to Cuba from Rome in the mid-1950s, fired up by their studies of Italian neo-Realism and full of enthusiasm for creating a Cuban film industry, rapidly found themselves imprisoned by the dictator Batista for their first short documentary. It was only after the Revolution that there was any real possibility of realizing a national cinema in Cuba. Thus it was a particular combination of individual technological expertise, acquired through a temporary migration, along with a specific set of political circumstances—a government committed to establishing autonomy for Cuba—that created the potential to effect a "definitive emancipation" through cinema. The mutual dependence of autonomy and authenticity is yet again confirmed. The case also illustrates how, in the ductile history of modernity, no particular factor has any inevitable value attached to it: the international context of the Cold War, which in the 1950s had acted as a constraint, had by the 1960s turned into an opportunity, and García Espinosa notes 1989 as a turning point (and mostly a negative one) in the history of Cuban cinema. He ends by emphasizing that the full emancipatory promise of film has not yet been realized because of the success of commercial filmmakers in replacing the "aura" of a work of art with the charisma of the film star. Suggesting that film festivals should award prizes to the "best character" rather than the "best actor," or the "least alienating film," García Espinosa develops a distinctively Latin American (which is specifically not national) version of modernity that retains and revives the original emancipatory promise of Enlightenment ideas, complemented by a commitment to overcoming the alienation dwelt upon by European modernists through social solidarity and ethical responsibility.

Néstor García Canclini opens chapter 9, "Culture and Communication in Inter-American Relations: The Current State of an Asymmetric Debate," by pointing out that many of the conventional debates about modernity in Latin America—exuberant modernism versus deficient modernization; the persistence of the traditional in a context of modernization—were all conceived within a national context. His main concern, now that the nation is no longer the main backdrop against which modernization occurs, is to analyze the effects on Latin America of a shift he identifies in recent decades from "Enlightenment modernity" to "neoliberal, globalised modernity." During the same period, the United States has displaced Europe as Latin America's main referent for modernity. In the context of these two phenomena, and building on earlier work (2002), García Canclini brings together analysis of socioeconomic transformations (particularly shifts in patterns of migration) and of social imaginaries. He argues that, while cultural exchange between "North" and "South" America does work both ways, the main point to emphasize is that it is asymmetric. This can be seen particularly clearly, he notes, in the contradiction between the U.S. embrace of social multiculturalism (affirmative action policies and so forth) and its concurrent marginalization of cultural goods—notably films—from outside its own territory. Thus multiculturalism—"the simple legitimacy offered by differences" (p. 188)—can function as a smokescreen. A crucial first step on the route to promoting the interconnectedness between the Americas, tolerance of difference and solidarity with subalterns that multiculturalists claim to seek, is to analyze the growing inequality created by the persistent asymmetrical power relations that ensure that the emancipatory promise of modernity is still not open to all. His analytical framework, seeing modernity in terms of the migration of people, goods, and ideas, is potentially applicable to earlier periods.

Divergences and Convergences

As will be evident from what has just been said, the contributors to this book take different positions on certain central problems of the topic: notably (1) the relationship between the objective and subjective aspects of modernity, and the related question of sources and their status; (2) the issue of whether modernity was imported, adapted, or invented and, if it came from abroad, whether it did so in successive waves or in one big bang (if so, which one?); (3) the role of ideas and intellectuals; and (4) the value of the term "modernity" in itself, either because of worries about its normative implications and/or because of concern that the category has to be bent so far in order to accommodate the varieties of experience that it had become meaningless and, therefore, analytically redundant. An emphasis on the constructedness and contestedness of modernity was enough to save the term for some (Radcliffe, Rowe), but not for others (Wade, Knight).

More unexpected is the degree of convergence on certain issues. First, there is broad consensus on the need to complement temporality with spatiality. Although those who focus on texts tend to think more about time, and those who focus on material culture tend to place more emphasis on space, all the contributors work on the basis that both should be taken into account. Sarah Radcliffe's reformulation of the question as "In which spaces was Latin America modern?"

won broad acceptance. Had the conference been entitled "Where was Latin America modern?" however, it is likely that "where" would have been interrogated as much as "when," and the need for a supplementary question along the lines of "in what domains?" or "in what spheres?" would have been identified. Second, all the contributors operate on the basis that the postcolonial paradigm is inadequate, especially in relation to Latin America, above all for its eternal return to the colonial encounter as the source of everything, including explanation. Even though historians of the region have tended to play down the changes brought about by the wars of independence (arguing for a periodization from 1750 to 1850), there is (as Radcliffe argues) a need for more work on the discontinuities between colony and independence, not least on the effects of the wars themselves in bringing about a fast-forwarding toward modernity (now an emerging area of historiography). Moreover, there is no automatic overlap between colonial/colonized and modern/non-modern or traditional, and indeed it is analytically crucial that all such dualisms be carefully historicized. Third, the two main avatars of modernization in the region, capitalism and the nation-state, were neither monolithic nor omnipotent. Indeed, as Radcliffe emphasized, drawing on David Harvey's work, capitalism operates precisely by exploiting existing variations in wealth. The nation-state is widely seen as a vehicle for modernity, but as much of the evidence presented here shows, cultural conceptions of the nation often won out over political ones. The emancipatory potential of the imagined community was thereby constrained as a gap opened up between a national ideal based on sovereignty and equality and the realities of arbitrary state power and dependency. As Claudio Lomnitz has argued elsewhere, the "resulting hybrids" have often been "interpreted as a manifestation of the resilience of national culture" so "the failure of modernizing projects is itself used to construct the national subject which is meant to be liberated by the nation-state" (Lomnitz 2000: 239). On the other hand, there is plenty of evidence, both in this volume and in the broader literature (Quijano 1990; Lomnitz 2001; Sáenz 2002), that the emancipatory promise of the modern has been remarkably persistent in Latin America.

Conclusion

The issues are not so much terminological as methodological. Some Latin Americanists argue for multiple modernities, arguing that such a framework allows for historicization as well as recognition of alternatives and challenges (Roniger and Waisman 2002; Whitehead 2006; and conclusion to this volume). Others resist what they see as an "easy pluralism" that "conceals" the history of imperialist domination and exploitation behind the constitution of modernist values (Sáenz 2002: viii). It is not easy in practice to shed assumptions about teleology and scaling, especially when they are built in to the emic usage of the concepts discussed. Moreover, as Radcliffe notes, modernity's power has operated historically precisely through its practices of privileging certain historical changes over others, and a strong analytical model of modernity would be able to take this into account. Although much useful work has been done on circuits of exchange of ideas and the reciprocity of constructions of self and other, in all of it there is a danger, as García Canclini reminds us, of eliding the enduring asymmetry of power relations between, say, the United States and any Latin American country. No single term or model in itself enables us to escape Eurocentrism, and anyway we all tend to classify those values we do not like as Eurocentric (often, individualism and free markets) and those we do (such as solidarity and autonomy) as subaltern. And often we take inordinate trouble to differentiate carefully in relation to Latin America whilst cavalierly dismissing the complexities of a "Europe" that has repeatedly produced strong internal critiques of its own values (as has the United States). The substitution of "multiple modernities" for "modernity" will not *in itself* secure the avoidance of determinism or condescension. In this collection, Rowe, whose chapter is the most radical in enacting, rather than merely proposing, a new methodology, retains the concept of modernity, albeit defined in the subjective, experiential terms that in themselves make it easier to accommodate difference. The "new analytical language" that is often called for (Sáenz 2002: x) is not enough on its own: it is also a matter of new questions, new sources and new perspectives, above all of preserving a constant state of awareness that outcomes could have been different, that processes interact with events in unexpected ways, and that language matters. That said, as Laurence Whitehead argues in his wide-ranging concluding chapter to this volume, the "multiple modernities" framework is potentially strong enough to accommodate these challenges.

In developing new approaches, we suggest that interdisciplinarity, for all its potential pitfalls, is not only desirable but fundamental. As discussions at the workshop showed, historical evidence (nowadays drawing on an increasingly wide range of sources) reminds us of the inadequacies of teleology. Literary and cultural studies' attention to language, textuality, and meaning draws attention to the aspirational aspects of modernity, to the recurrent idea that the modern is always elsewhere (or, if the argument is taken to its extreme, as in Bruno Latour's *We have Never Been Modern*, nowhere). In other words, modernity is best seen not as an achieved state, but as endlessly deferred by definition. Literary history also highlights that the modern by no means always moves from centre to periphery: after the First World War, "avant-garde movements appeared simultaneously in the margins and in the center" (Geist and Monleón 1999: xxx). The social sciences compel us to attend to the implications of the analytical terms we choose.

Thus, in thinking about the role of external models in the creation of Latin American modernity, it is possible to see European experiences as historical precedents without necessarily seeing them as normative. Historically, a variety of experiences not only from Europe and the United States but also from many other places (notably Japan, China, the Soviet Union, Australia, and New Zealand) did indeed function as guides in Latin America—although often as to what *not* to do. Work on the historical reconstruction of how external ideas were assimilated, adapted, challenged, and appropriated in Latin America has supplied ample evidence that in itself challenges normative assumptions, although there is far more to be done in this area. It was not always the case that outside models were well received in Latin America and contested only by difficulties of implementation. At least from the early twentieth century onward, critiques and alternatives were proposed from within the region, responses that cannot be adequately understood if conceived in terms of resistance to the modern. Although it seems to be the case that the term "modernidad" only became current in Latin America in the context of recent debates about postmodernity, "moderno" and "lo moderno" was certainly used, for example, in the titles of popular magazines, from the early twentieth century onward. The key question is how external models were mediated, which was far more varied than has always been acknowledged.

In this respect, the way forward seems to lie in an analytical approach that enables us to leave behind the argument about whether ideas or material forces are the prime agents of history. Surely, as Charles Taylor has argued, human practices are always to some extent, even when "coercively maintained," based on "self-conceptions, modes of understanding," whereas "ideas always come in history wrapped up in certain practices, even if these are only discursive practices" (Taylor 2004: 31 and 33). His model of the social imaginary, or "what enables, through making sense of, the practices of a society" (Taylor: 2), is one proposal for going beyond this false dichotomy, and also offers a way of giving social depth to a topic that is all too often analyzed in relation to elites. Radcliffe's conception of modernity in terms of discourse, project, and experience might usefully be extended to include historical consciousness, which would allow for the incorporation of the argument that modernity entails a particular consciousness of time, denoted especially by anticipation of a progressive future and a sense of accelerating change (Koselleck 2002). In any case, it seems important to find a model of modernity that can incorporate both objective and subjective elements, rather than seeing it either as an outcome of measurable historical processes or as a nebulous cultural project. Modernity is perhaps best seen as a cluster concept, as a set of aspirations and potentialities, any of which can be emphasized, reinterpreted, criticized, celebrated, or marginalized in any particular historical situation, but none of which can be entirely discarded if a state of "modernity" is to command widespread recognition. In sum, we suggest that both the humanities and the social sciences need to find room for the kind of question that is not cognate either with value-laden terms such as "happy" and "good" (chapter 4 by Alan Knight, p. 91) or with the more readily verifiable "literate," "urban," or "industrial": a question like, for all its flaws, "when was Latin America modern?"

References

Barthes, Roland (1977). *Image-Music-Text*, Essays selected and translated by Stephen Heath. London: Fontana.

Bayly, C. A. (2003). *The Birth of the Modern World 1780–1914.* Oxford: Blackwell.

Brunner, José Joaquín (1992). *América Latina: cultura y modernidad.* Mexico: Editorial Grijalbo.

Dussel, Enrique (1995). *The Invention of the Americas: Eclipse of "the Other" and the Myth of Modernity.* Trans. Michael D. Barber. New York: Continuum.

Forment, Carlos A. (2003). *Democracy in Latin America 1760–1900. Vol. I. Civic Selfhood and Public Life in Mexico and Peru.* Chicago and London: The University of Chicago Press.

García Canclini, Néstor (1992). *Culturas híbridas. Estrategias para entrar y salir de la modernidad,* Editorial Sudamericana, Buenos Aires. In English as *Hybrid Cultures: Strategies for Entering and Leaving Modernity,* trans. Christopher L. Chiappari and Silvia L. López (1995). Minneapolis and London: University of Minnesota Press.

——— (1993). *Transforming Modernity: Popular Culture in Mexico.* Trans. Lidia Lozano. Texas: University of Texas Press.

García Canclini , Néstor (2001). *Consumers and Citizens: Globalization and Multicultural Conflicts.* Trans. George Yúdice. Minneapolis and London: University of Minnesota Press.

——— (2002). *Latinoamericanos buscando lugar en este siglo,* Buenos Aires: Editorial Paidós.

Geist, Anthony L. and José B. Monleón, eds. (1999). *Modernism and Its Margins: Reinscribing Cultural Modernity from Spain and Latin America.* New York and London: Garland Publishing, Inc.

Geras, Norman and Robert Wokler, eds. (2000). *The Enlightenment and Modernity.* Basingstoke: Macmillan.

Guerra, François-Xavier (1992). *Modernidad e independencias: ensayos sobre las revoluciones hispánicas.* Madrid: Editorial Mapfre.

——— (1995). *Revoluciones hispánicas: Independencia americana y liberalismo español.* Madrid: Editorial Complutense.

Hardt, M. and Negri, T. (2000). *Empire.* London: Harvard University Press.

Hopenhayn, Martín (1994). *Ni apocalípticos ni integrados. Aventuras de la modernidad en América Latina,* Fondo de Cultura Económica. In English as *No Apocalypse, No Integration: Modernity and Postmodernism in Latin America,* trans. Cynthia Margarita Tompkins and Elizabeth Rosa Horan (2002). Durham: Duke University Press.

Jrade, Cathy L. (1998). *Modernismo, Modernity, and the Development of Spanish American Literature.* Austin, TX: University of Texas Press.

Koselleck, Reinhart (2002). "The Eighteenth Century as the Beginning of Modernity," in Koselleck, *The Practice of Conceptual History, Timing History, Spacing Concepts.* Trans. Todd Samuel Presner et al. Stanford, CA: Stanford University, 154–69.

Lomnitz, Claudio (2000). "Passion and Banality in Mexican History: The Presidential Persona," in Luis Roniger and Tamar Herzog, eds., *The Collective and the Public in Latin America: Cultural Identities and Political Order.* Brighton and Portland, OR: Sussex Academic Press, pp. 238–56.

Lomnitz, Claudio (2001). *Deep Mexico, Silent Mexico: An Anthropology of Nationalism.* Minneapolis and London: University of Minnesota Press.

Pred, Allan and Watts, Michael (1992). *Reworking Modernities: Capitalism and Symbolic Discontent.* New Brunswick: Rutgers University Press.

Quijano, Anibal (1990). *Modernidad, identidad y utopía en América Latina.* Quito: Editorial El Conejo.

Rama, Angel (1984). *La ciudad letrada.* Hanover, NH: Ediciones del Norte. In English as *The Lettered City,* trans. John Charles Chasteen (1996). Durham, NC: Duke University Press.

Ramos, Julio (1989). *Desencuentros de la modernidad en América Latina: literatura y política en el siglo XIX,* Fondo de Cultura Económica, Mexico. In English as *Divergent Modernities: Culture and Politics in Nineteenth-Century Latin America,* trans. John D. Blanco (2001). Durham: Duke University Press.

Rincón, Carlos (1995). *La no simultaneidad de lo simultáneo: posmodernidad, globalización y culturas en América Latina*: Santafé de Bogotá: Editorial Universidad Nacional de Colombia.

Roniger, Luis, and Carlos H. Waisman, eds. (2002). *Globality and Multiple Modernities: Comparative North American and Latin American Perspectives*. Brighton: Sussex University Press.

Sáenz, Mario, ed. (2002). *Latin American Perspectives on Globalization: Ethics, Politics and Alternative Visions*. Lanham, Boulder, New York and Oxford: Rowman and Littlefield Publishers.

Taylor, Charles (2004). *Modern Social Imaginaries*. Durham: Duke University Press.

Thurner, Mark (2003). *After Spanish Rule: Postcolonial Predicaments of the Americas*. Durham and London: Duke University Press.

Uribe-Uran, Victor M., ed. (2001). *State and Society in Spanish America during the Age of Revolution*. Wilmington, DE: Scholarly Resources, Wilmington.

Whitehead, Laurence (2006). *Latin America: A New Interpretation*. New York: Palgrave.

Part I

Views from the Historical and Social Sciences

Chapter 1

Geographies of Modernity in Latin America: Uneven and Contested Development*

Sarah A. Radcliffe

Introduction

Modernity as it is most commonly understood comprises a constellation of knowledge, power, and social practices that first emerged in Europe in the sixteenth and seventeenth centuries and slowly extended over space (Gregory 1994: 388–92). This definition contains two components making it instantly recognizable: namely, its Eurocentrism and its underlying model of spatial diffusion from a core "modern" area. In other words, our concepts of modernity contain within them a distinctively *geographical* perspective, a mental map for reading the world and the position of Latin America within it. A critical reading of this underlying spatial language of modernity engages us in an analysis of Eurocentricism in our *theory* and the spatial processes behind modernity's *materiality*. In this context, my chapter attempts to extend existing critiques of the framing of modernity (e.g., Mignolo 2000), while additionally asking us to push our understandings of modernity further in order to provide a more updated spatial theory than diffusionism.

Instead of asking about the temporality of Latin American modernity, then, I rework the question to ask *in which spaces* was Latin America modern? My chapter attempts to demonstrate that Latin America was not a subordinate and peripheral latecomer to the modernity show. This chapter explores some aspects of geography's reasoning about modernity, drawing on substantive material from Ecuador. First, I provide a critical overview of several geographical approaches to modernity that highlight both the importance of a nondiffusionist theory and the significance of geographical practices in the construction of modernity. The second section outlines a tentative framework for an understanding of Latin America's modernity. In the third section, I explore how we might apply a theory of the geographical practices underpinning modernity, using material on the Ecuadorian nation-state, and development, before giving a few concluding thoughts.

Geographical Perspectives on Modernity

Within the discipline of geography, there are broadly speaking two subgroups of geographers working on modernity, namely historical-cultural geographers and development geographers. Historical-cultural geographers generally tend to ask questions about the nature of early metropolitan modernities, whereas development geographers are interested in the global South. Although in the mid-twentieth century, geography was often preoccupied by the *lack* of modernity in the developing world, today there is more of a postcolonial sensibility to what has been termed multiple reworked modernities (see later). Overall, there is an underlying fascination with those geographical areas/historical periods where modernity cannot be taken for granted, and where its self-proclaimed self-evidentiality can be placed under close scrutiny.[1] The discipline has generally approached modernity when it is less than secure, either at its initiation in "core" countries, or where underdevelopment and neocolonialism has undermined it. Both development and historical theorists draw on the insight that society and space are constituted simultaneously and interdependently (Agnew 1987; Pred 1984). Drawing on these analytical traditions, this section outlines key human geography concepts and approaches and illustrates their arguments by brief reference to Latin American examples. By means of a critical engagement with existing concepts and approaches, I work toward a framework for the Latin American case.

In order to avoid "both ambiguity and totalisation" (Ogborn 1998: 2), definitions of modernity have to grapple with its projections into the future, powerful discourses and representations that work to establish modernity's reference points, as well as its embodied experiences and identities. In this sense, modernity can be viewed as a three-fold phenomenon comprising a project (or projects), discourses, and experiences (Jervis 1998; Howell forthcoming):[2]

   (a) A *project*—Modernity defines itself in relation to the future, and invokes planning, projects, and regularization, as well as administration and bureaucracy, an approach associated with Foucault and

Habermas. Nevertheless, as a project, modernity is associated with cumulative improvement that cannot be relativized. For example, declining child mortality rates resulting from vaccinations and health care is preferable to children dying. Moreover, Enlightenment ideals of banishing ignorance, misery, and despotism are considered to be core to the project of modernity. Modernity comprises a spatial as well as a social project, in which the categories, orders, and regularizations by which modernity is recognized are crucially and simultaneously spatial categories, orders, and regularizations (Foucault 1984; Bauman 1991). "[Boundaries must be] sharp and clearly marked, which means 'excluding the middle,' suppressing or exterminating everything ambiguous, everything that . . . comprises the vital distinction between *inside* and *outside*" (Bauman 1991: 24, original emphasis). These goals informed development thinking throughout the twentieth century in LA, where—as in other parts of the world—modernity was a project centered on urban spaces (Martins and Abreu 2001; Popke and Ballard 2004).

(b) A *discourse*—Modernity as a discourse can be viewed as claims about superior, non-"traditional" practices, ideologies, or projects in which (supposedly universal) improvements are not always readily apparent in practice. Discourses of modernity are explained by reference to the cultural hegemony and violence that differentiate between that which ranks as "modern" and the "non-modern." This discourse constitutes the doubleness of coloniality/modernity as the simultaneous and violently established hierarchies of modernity from the discovery of the Americas (Mignolo 2000; also Lechner 1995). In Andean countries, for example, modernizing agrarian development through most of the twentieth century was founded on discourses that large farms were more efficient than small peasant plots. Discursive aspects of modernity place much emphasis on what is new (technology, art, forms of social organization, economic practices), and dismiss what is traditional or old, seen as superstition or ignorance, the myths of religion.[3] The constant shifting of modernity's parameters and content does not, however, lessen its power to raise the specter of failure in "keeping up" with modernity. In Bolivian neoliberal modernity, the crucial differentiation becomes "changes in the mentality of Bolivians who should favour a more productivist attitude, be more reliant on their own efforts and less on state protection, a more open attitude to exchange with the outside world and less *ukhu runa*" (Oporto 1992: 86).[4] The discourse of modernity thus provides no closure, as it is far from hegemonic and must work to align its projects and events with the judgments and evaluations of multiply positioned subjects, a point that brings us to experience.

(c) An *experience*—Modernity would be nothing if it were not experienced, embodied, and performed by individuals in multifarious ways. Ordinary men and women experience—and are molded by—changes and discourses in ways that constitute the projects and discourses of modernity. The constitution of modernity rests upon the instantiation of certain behaviors, deportment, and embodied responses in social subjects (Hansen and Stepputat 2005). Giddens (1990) argues that one of the key features of modernity is reflexivity, a constant self-critical examination of social practices and their modification which, he argues, contributes to the volatility of the modern, to "radicalised modernity." For example, gender roles in Latin America through the twentieth century were radically transformed by projects and discourses of modernity that constituted new roles for women and men (as workers, participants in the market economy, and political subjects). As they gained economic, political, and social rights, women's experience of family life, work, relation to the public sphere and the polity, and Enlightenment goals were reworked, albeit not in a simple trajectory toward greater emancipation or self-realization (Dore and Molyneux 1999).

(d) By encompassing projects, discourses, and experiences, the depth and complexity of modernity can be retained, permitting us to turn now to outlining how geographers have addressed these facets of modernity. By means of a critical reading of geographical theory, the chapter then moves toward a potential framework for understanding Latin America's modernity.

Projects

Geographers have identified either capitalism or political power—expressed in the form of the modern nation-state—as the driving force behind modernity's projects and as instantiating modernity's spatiality.[5]

Capitalism is and always has been a global process, unifying silver mining by Andean Indians and the labor of transported African slaves into complex lines of connection, which contributed to globally uneven distributions of wealth. In this context of highly diverse and interconnected political economies, the West managed to imagine itself as modern and—later—as developed. As James Mahoney (2003) shows, the insertion of different regions of Latin America into global political economies has long-run impacts on incomes, life expectancy, and literacy rates. Yet to view political economy as determinant can also be problematic. First, when political economies are reduced to capitalism then accounts of modernity become accounts of the all-powerful machinations of a hegemonic and abstract capitalist force (Harvey 1989; compare Ogborn 1998). Second, the sheer diversity of forms of political economy—even in Latin America—suggests that capitalism's diversity and responsiveness to local/regional conditions has to be accounted for (e.g., Guano 2002). Third, Marxist accounts of political economy risk prejudging the outcomes of capitalist (under)development, whereas accounts that stress the contingency, nondeterminacy, and non-teleological nature of regionally specific political economies are more accurate.

Accounts of uneven development overcome some of these potential limitations by stressing the inherently unevenly expressed nature of capitalist economies and indeed the basis of political economic dynamism in unevenly developed landscapes. As noted by geographer Neil Smith (1984), "The uneven development of capitalism is structural rather than statistical . . . uneven development is the systematic geographical expression of capital." According to Smith, capitalism works over existing surfaces of economic activities—mines, plantations, industrial districts, service providers—seeking out areas for profit, where it can make over activities thereby generating a new landscape over which, again, capital seeks out the means for another round of "creative destruction" (Harvey 1989). However, such an account entails its own Eurocentricity as it relegates areas of peasant agriculture, informal production, and non-monetized social reproduction to an undifferentiated area of potential capitalist activity, thereby denying them agency and *sui generis* uneven landscapes. In this context, development geography has focused on what geographers Allan Pred and Michael Watts term "re-worked modernities." In their analysis of diverse locales across the world (except Latin America), Pred and Watts highlight how "difference, connectedness and structure are produced and reproduced within some sort of contradictory global [capitalist] system" (Pred and Watts 1992: 2). While attending to local specifics of modernity "experienced, constituted and mediated locally," they place this within the frame of "non-local processes driving capital mobility" including noncapitalist forms of economy and nonmarket sociocultural relations (6). Highlighting the mutual embeddedness of culture, economics, and politics and actors' agency permits recognition of multiple forms of modernity in the global South. In other words, the concept of reworked modernities "recasts metropolitan modernity as its dominant form rather than as its universal standard" (Coronil 1997: 9).

As culture and politics are attributed significant weight in shaping the localized/regionalized clusters of production, reproduction, and circulation, Latin America can be viewed as the location of multiple reworked modernities. In Latin America today then, neoliberal political economies are not determinant in its modernity as regionally specific political economies show contingency, nondeterminacy, and non-teleological outcomes (e.g., Guano 2002; Perreault and Martin 2005; Radcliffe 2005a). Neoliberalism holds out the promise of reworking the uneven, undeveloped regions of Latin America to generate wealth and modernity, although it is as prone to (radical) unevenness and failed projects as previous development models.

In relation to politics, geographers have considered the power relations of the modern nation-state as fundamental to the nature of modernity's projects. As in other disciplines, the administrative practices of the modern state have often been identified as key to the nature of modernity. As the locus of power in modernity, the nation-state has played a central role in the disposition of subjects, landscapes, and resources in a "rational," self-evident topography. As modernity's projects have to be put into practice, the nation-state is at the center to establish regularities and mappings of objects and subjects. Such a project is founded upon the designation of spaces and interactions. The modern nation-state orders the regular disposition of subjects, resources, and landscapes. State power in turn rests upon three pillars, each of which is intrinsically spatial—the sovereignty of clearly bounded territorial spaces; an assumption of fundamental opposition between domestic and foreign affairs; and the actions of a territorial state as a geographical container of modern society (Agnew 1999: 175–6). Modernity's landscape of power is hence a spatial disposition, a story about the location of objects, boundaries, and subjects and their interrelationships across space (as the world does not rest on the head of a pin).

In this context, simple cartographic practices of mapping and inventories of territory not only encompass information in a particular graphic form but they also function to uphold modern forms of power. The spatiality of modern territorial power is thus supported by specific *geographical practices* through which territories are known and contain power, and by means of which the self-evidentiality of modernity can be reiterated. Accordingly, maps—and the notions of sovereignty, power, and international relations that they convey— underpin the performance of sovereignty that is the modern attribute of statehood. Discussing the case of war-torn Afghanistan, political geographer Simon Dalby argues:

> In the places . . . designated Afghanistan, despite the absence of most
> of the normal attributes of statehood, . . . the continuing fixation on a

simple cartographic description in circumstances where sovereignty has
to be intensely simulated to render categories of political action meaningful, highlights the centrality of geographical practices to modernity.
(Dalby 2003)

Whether in Mexico, Ecuador, or Afghanistan, the status as modern
rests upon the iteration of a territorial claim, a sovereignty and projected representation of identity through which place-specific claims
to modernity are produced. Latin American modernity—and other
modernities around the globe—is bound up in a "territorial trap"
(Agnew 1999) in which the practices of mapping, definition of borders, and inventories of landscape are part and parcel of modernity's
project. The production, circulation, and skilled reading of maps and
geographical information are not an innocent or incidental activity in
the configuration of modernity, as the case of Ecuador illustrates later.

* * *

Overall, geographers have focused attention on the experiences of space
and the transformation of spaces, places, and landscapes. Spaces of
modernity are "quite simply, a series of 'multiple and contradictory'
spaces and places at all scales and taking many different forms" (Ogborn
1998: 20), thereby going beyond a view of modernity's spaces as exclu-
sively rational, panoptic, and geometric. In contrast to Henri Lefebvre's
account of the capitalist production of modern abstract space as
planned and rational—"the fantasy of the straight line" (Pred and Watts
1992: 16; Lefebvre 1991)—the emphasis is on the "differentiated and
contextualised geographies of modernity's spaces . . . [and their] contours and conditions of existence" (Ogborn 1998: 21).[6] By examining
the projects of modernity as inherently spatial, geographers have highlighted the contested hegemony of reworked political economies and
the modern state in providing a grid in relation to which modernity is
stretched out across the globe in highly uneven and incommensurate
topographies. Economies and political power are expressions of power,
whose operation cannot be removed from the heart of modernity: place
in the double sense of location and rank/order is thus central to accounts
of Latin American modernity (Robinson 1989: 176). The projects of
modernity founded on political economy and the state are only constituted in as much as they relate to specific practices of (re)production
and exchange, and to particular geographical practices through which
territorial sovereignty is enacted. Not a mere backdrop to the "real
business" of modernity, the organization of spaces is central to the very
notion and infrastructure of modernity.

Discourses

In contrast to the aforementioned accounts, historical and cultural geographers emphasize the discursive and cultural political dimensions of modernity, attributing representational hierarchies with primacy in accounts of modernity's emergence and development. Considerable attention has been paid to colonial discourse and representation and constructions of modern status in colonial powers, and designation of colonized areas as non-modern.[7] Geographers are coming around to an analytical framework in which, as the cultural geographer Derek Gregory puts it, the "viral presence" of colonialism was "always as much about making other people's geographies as it was about making other people's histories" (Gregory 2004: 7, 11). In discussing the colonial modernity of today's Middle East, Gregory highlights the ways in which these geographies are encountered and produced where culture meets power. Gregory resists reducing modernity to an account of capitalism's uneven development and instead stresses what Edward Said usefully termed "colonialism's imaginative geographies" (Said 1978). Paraphrasing Said, Gregory defines imaginative geographies as

Partitions . . . that serve to demarcate "the same" from "the other," at once constructing and calibrating a gap between the two . . . a familiar space that is "ours" and an unfamiliar space beyond "ours" that is "theirs" . . . [Imaginative geographies] are fabrications, a word that usefully combines something fictionalised and something made real, because they are imaginations given substance. (2004: 17)

In addition to colonials' Orientalist images of subalterns,[8] the Janus-faced nature of imaginative geographies requires an enquiry into Occidentalism, defined by Fernando Coronil as "representational practices whose effect is to present the non-Western peoples as others of a Western self" (Coronil 1997: xi, 13–5; also Mignolo 2000).[9] In other words, analysis of the colonial modern has led geographers to explore how relationships between places have been constituted around a geographical imagination and around the metropolitan self-fashioning in opposition to its (subaltern) Others. From their different area specialisms, geographers have begun to deconstruct the notion of a single or unitary colonial power and postcolonial histories (Power 2003). However despite deconstruction of the singularity of colonial experiences, there is a surprising lack of substantive work by geographers on the construction of Latin American imaginative geographies under colonialism, although see Harley (1992) and Scott (2003).

However, as in postcolonial studies, the risk here is to fold every explanation back on to the colonial encounter, failing to recognize the diversity of colonial forms of power and mutual construction, resistances to colonialism, and the flexibility and noncomparability of postcolonial experience (Said 1978; Young 1990). Rather than treat colonial and republican periods as colonialism Parts I and II, more needs to be done on *dis*continuities. Beyond the commonplace criticism that postcolonial accounts reduce materialities, practice, and contradictory subjects to the representational, there is also a problem with their *geography*. By reinscribing the colonial difference—that is, the distance and boundaries between colonial power and colonized territories—such accounts risk being blind to other geographies of difference and imaginative geographies. This point has special relevance for Latin America, where a postcolonial approach downplays the imaginative geographies and territorial interventions constructed in the nearly two centuries of republican rule, by collapsing down this history onto an outdated singular boundary.[10] Mignolo thus treats the Enlightenment as derivative in Latin American modernity, coming as it did well after the instantiation of colonial modernity, or coloniality was "quite simply, the reverse and unavoidable side of 'modernity'—its darker side" (2000: 19, 22).

Colonial/colonized boundaries comprise not an enduring marker of modernity/non-modernity, but a shifting topography of power and difference. A more dynamic approach to (post)colonial difference views discursive and cultural relationships between areas as contingent and constructed, malleable and subject to sudden reversals. Following Enrique Dussel (1993, 2002), we can agree that it was precisely Spain's conjuncture of defeating the Moors then immediately expanding its colonial enterprise onto the New World that permitted Spain/Europe to consider itself as the fulcrum, the center of a new epoch of modernity.[11] Similarly, Ogborn views eighteenth-century London as "the hinge to the modern world" (1998: 32; also Gregory 2004). The metaphor of the fulcrum—a delicate balance between places in a maelstrom of projects and discursive negotiations— provides a means of recognizing the contingent and constantly shifting geographies of modernity, rather than fixing them in the bounds of a formally established colonial regime's rigid frontiers, or indeed its republican replacement.

Rather than taking (modern/non-modern) areal differentiation for granted, this approach interrogates the (discursive, material, embodied) means by which areas remain *connected* and creates the context for a "fulcrum," a moment of contingent and non-innocent contact.

From a postcolonial perspective, geographers have begun to understand the role of connection and mobility in the global South's distinctive modernities. As development geographer Marcus Power
argues, "Postcolonialism is partly about thinking through the implications of *stretched-out geographies*, making connections and understanding the important flows and movements between North and South"
(Power 2003: 122). These "stretched out geographies" have historically and regionally specific topographies; Africa cannot be mapped
out on to the grid of these stretched out geographies in the same way
as Latin America. Moreover, Ecuador cannot be positioned on this
topography of connections in the same way as Mexico or Argentina.
Geographies of modernity are simultaneously about connection *and*
difference, hierarchy *and* interrelations (Pred and Watts 1992: 13).
Looking at complex, (non)colonial itineraries of connections permits
us to examine how development interventions and colonial relations
are embedded in geographies of political economies and cultures.
Whereas many of these stretched out geographies were oriented
toward Europe in the nineteenth and early twentieth century, North
America is now a common locus of these connections (García Canclini
2000: 211; Slater 2004).[12] These topographies may owe something to
colonialism but it is insufficient to attribute all of their contemporary
problems to colonialism or a closed category of Western development.

Experiences

Finally, accounts of modernity must also examine the question of
agency, the body and the performativity of modernity. Whereas classic
accounts of modernity stress the intellect/the mind as the prime site
for the achievement of Enlightenment modernity, post-structuralist
accounts stress how modern practices/projects often work through
the performance and embodiment of values, subjectivities, and moral-
ities. In the metropole and the colony, the disciplining of bodies as
bearers of a modern future has, following Foucault, been taken as a
fundamental part of modernity. Cultural geography makes a double
move from this approach, arguing that the instantiation of disciplined
bodies occurs through a disposition of space (prison, asylum, and so
on) and that local spaces are nested within other spatial scales.
Gregory argues: "European modernity constructed the self—as the
sane, the rational, the normal—through the proliferation of spacings.
But these were all spacings within Europe" (Gregory 2004: 3).

Gender, race, and colonial difference underpinned the construction of subjects and identities within a world of colonial difference

(Stoler 1995). Although the discourse of modernity often prioritizes racial or cultural homogeneity, the ever-changing proliferation of different racializations and their attendant advantages vis-à-vis modernity means that embodied racial meanings underscore modernity (Korff 2001; Bonnett 2002).

Self-ascription as modern rests upon the alignment of multiple scales around consistent markers of modernity: the body, the home, the city, the nation, the world region. The multiple scales/spacings of embodied and experiential modernity are thus mutually interactive, having a relational influence on the constitution of modernity. In the words of Miles Ogborn, "Modernity has both created and confounded spatial scales" (Ogborn 1998: 19). Practices of modern power are constituted in diverse geographies—material and imagined— around which discursive and practical alignment is constantly attempted. The scale of individual bodies thus links to the wider organization of spaces. In colonial Latin America, the use of a chequer-board street system around a central plaza and the forced relocation of indigenous populations to lower altitude settlements illustrate how internalization of modernity links to specific landscapes of power (Robinson 1989), while in twentieth-century Bolivia, "Modernity transformed gendered and racialised bodies into sites of conflict" (Stephenson 1999: 3). Embodiments are the outcome and the effect of a disciplining process that results from the coordination of space, place, and society by groups invested with power. Space/power are entailed in the very process of embodiment.

The iteration of discipline operates together with slippages and everyday forms of resistance that produce embodiments of subjects whose spacings are not aligned to spaces of disciplinary power. The possibility remains of practices at diverse scales that disrupt the modern spatial order in ways that cannot create simple romantic anti-moderns. Various dimensions of this resistance can be mentioned. First, the subalterns' capture of techniques to order space in order to reconfigure the project of modernity, for example, indigenous organizations using maps to question the modernizing development plans of Peru or Ecuador (Orlove 1993; Radcliffe 1996). Alternatively, the maps of modernity can be deliberately misread by actors (Orlove 1991; Radcliffe and Westwood 1996), or actors can engage with modern organization of space in ways that are incommensurate with—or slightly reorient away from— mainstream projects of modernity. For example, stretched out geographies of connection link urbanward and international migrants and Andean villages to fund village festivals and hybrid cultural forms (also Oslender 2004). In other

words, multiple resistances to—and the non-hegemonic incompleteness of—modernity produces not antimodern or postmodern subjects
but diverse positionalities vis-à-vis uneven development and topographies of power and difference. The geographically variable benefits
and costs of modernity shape efforts to restructure local/regional/
global forms of modernity. Social movements for rights arise from and
reflect the (physical, social, and political) distance of groups and/or
areas from the state (Davis 1999; Slater 1998). Latin American social
movements as diverse as the Zapatistas and Afro-Colombians devise
political strategies that are simultaneously spatial tactics, reworking
the spaces of modernity in line with their own priorities (see Slater
2004; De la Fuente 2004).

Non-Eurocentric Spaces of Modernity

To summarize, modernity comprises projects, discourses, and experiences, each constituted by and given expression in uneven topographies, interconnections, and diverse geographical and disciplinary
practices. Whereas the project of modernity is largely a global story
(capitalism, Western science, Enlightenment), the discourses and
experiences of modernity striate into regional, postcolonial and disaggregated spaces and boundaries. In the context of uneven development and connections across space, modernities can be recognized as
multiple and incommensurate once the dualistic imaginative geographies of postcolonialism are discarded. The dynamism of uneven development combined with the incompleteness of modernity's projects
and diverse resistances together result in highly diversified geographies
and embodiments of modernity.

It is useful here to distinguish between the spatiality of modernity
and geographical practices. The spatiality of modernity comprises the
underlying spatial processes that shape—literally—the map of modernity and whose dynamics undergird the distinction between Latin
America and the rest of the world and between different subregions
and countries within Latin America. This point is worth reiterating as
it counters aspatial accounts of modernity. Although providing
insights into the contemporary workings of power and resistances to
them, Bauman (1999) and Hardt and Negri (2000) argue that power
is now extra-territorial and hence the project of modernity is somehow
freed from geography, primarily from the coordinates of the modern
nation-state. By contrast, I am suggesting that different phases and sections of the uneven spaces of modernity have different spatialities; they
are always spatial although not always state territorial. In other words,

it is impossible to disentangle multiple modernities from their spatial underpinnings. Drawing on the notion of reworked modernities, colonial modernity and the multiple forms of "amazing capitalism," I argue that the spatialities of modernity create highly differentiated maps of modernity across the world. Subject to the dynamism of uneven political economies—not all of them capitalist—and the power relations embedded in colonial/postcolonial imaginative geographies, areas of Latin America experience highly diverse forms of modernity. None of these diverse forms map out exactly onto the experiences of Europe or North America, precisely because of the intrinsically *geographically variable* nature of these economic, cultural, and political processes, or "historical experiences of multiple local histories (the histories of modernity/coloniality)" (Mignolo 2000: 22). Keeping in sight the relational and "stretched out" connections between Latin America and the rest of the world permits an understanding of modernity's spatiality. This spatiality, to reiterate, is not reducible to the diffusionist metaphor of classic definitions of modernity.

Within these macro-scale transformations of modernity, geographical practices play a vital role in the construction of modernity's practices, projects, and experiences. These include the ideological and practical work of cartography, mapping, territorial imaginations, creating homogeneity over an extension of space, and creating narratives of place and locale. Each of these contributes to the spatial ordering of modernity.[13] Yet these components of modernity's practices do not all operate over the same extensions, but are associated with different scales, such as an urban development project or a nation-wide mapping of cocaine production. Modernity's spatial topography is constituted across scales. From the body to the city to the nation-state and empires, different scales interact and see power move across from one scale to another in the constitution and understandings of modernity. Given these incommensurate extensions of modernity's practices, discourses, and experiences, we come to recognize the ways in which modernity operates in different ways across different scales, expressing its power and impacts differently on the human subject and the national territory (and other scales between these). Foucault provided important insights into how power operates across from the economy, through the polity and into social institutions. Recently, geographers have discussed how the spatiality of social life produces different, interconnected, scales, which in turn structure certain outcomes and prevent others (Marston 2000). Whether seen as a project, a discourse, or an experience, modernity cannot be reduced to a straightforward process of construction, and moreover it remains a highly contested process. Struggles over modernity arise from the slippage between projects, uneven development, and contestations.

Latin American Modernities in Spatial Perspective

Nation-States

"Order and Progress" (motto on Brazilian flag)

Although reflecting the derivative moment of Latin American coloniality/modernity (Mignolo 2000: 19), the nation-state in the region has often been understood to comprise both an agent and symbol of modernity. Its organizational logics, forms of administration, philosophies and practices of citizenship, inventories, and forms of rule and ruling, each contribute to the project, experiences, and discourses of modernity. Statehood establishes a territory as self-evidently sovereign, and comprises an entity greater than the sum of its parts. Postcolonial states can be analyzed in terms of "how various languages of stateness, not necessarily all purely Western in origin, have been spread, combined and vernacularised in various parts of the world" (Hansen and Stepputat 2001: 10). In other words, stateness/statehood is taken to be a project, one that moreover reflects the specifics of the terrain, culture, and society over which the state exerts sovereignty. Drawing on Joseph and Nugent's influential study of nation building in Latin America, we can suggest that there are "practical and processual dimensions of state formation" through which modernity's project can take shape and around which resistances and reworkings occur (Joseph and Nugent 1994: 19).

The underpinning of state projects of modernity by means of geographical tools, knowledges, and imaginations is illustrated here in relation to the Ecuadorian case. As I have argued previously, postcolonial Ecuadorian statehood "is a story of the deployment of geographical terms and the selective adoption of ideas of spatial science" (Radcliffe 2001: 124).[14] These methods include cartography, inventory, and census organization, as well as the physical integration of territories by such means as currency, transport, and education.[15] Another way of viewing these practices—along with their attendant discourses, and interpellations of subjectivity—is to see them as technologies of spatial power, tools for the consolidation of territorial sovereignty, and means of social integration in short, to see modernity as a move toward order and coherence. Not a detailed historical account, this represents an outline approach to statehood in terms of modernity's spatiality. Two key features of Ecuadorian modernities are its basis in geographical organization and the imaginative geographies, and their constant readjustment and realignment in line with shifting notions of the "modern." In the historical geographies of Ecuadorian state modernity, four moments can be distinguished: 1860–1875, 1930s–mid-1940s, the 1960s and 1970s, and the 1980s and 1990s.

Making the Geographic Tools of Statehood Scientific: 1860–1875

Modern postcolonial statehood in the sense of sovereign territoriality in Ecuador was not a foregone conclusion by the mid-nineteenth century: "The national question in 1860–1865 was the territorial question" (Quintero and Silva 1994: 114). President García Moreno turned to the scientific community in Europe, following a Latin American trend to engage with positivism, observation, and experiment (Hale 1996: 148). Geographers, along with botanists, naturalists, and mathematicians, were employed in the newly established Escuela Nacional Politécnica (the first non-church university), many of them Europeans. One of these, Theodore Wolf, a German geologist arriving in 1870, together with Wilhelm Reiss and other colleagues, mapped the country, thereby establishing centralized knowledge about its layout and makeup. Cartography was a prime new tool, and assisted the reworking of subregions and hence the political balance of power. By 1870, the state had created the post of state geologist, held initially by Wolf, and inventories of mineral resources combined with geographers' growing professionalization to shape the nature of public works administration.[16]

The practical integration of the territory was complemented by an emergent geographical imagination and system of representation. Grasping a vision of a unified territory—as seen earlier, a characteristic of modernity's territorial trap—President García Moreno oversaw the unprecedented building of bridges, roads, and the Guayaquil-Quito railway line (the latter not completed until 1908), although the coverage remained in practice extremely sketchy.[17] The nationalist agenda behind the Quito-Guayaquil railway was unusual in the Latin American context (Foote 2004: 89). The homogeneous space of currency circulation began with the sucre's issue in 1884, while the national anthem of 1866 provided the glimmerings of an imagined community. It is also perhaps interesting that the anthem's author, Juan León Mera, also wrote *Catechism of Geography of the Republic of Ecuador* (Terán 1983: 183).

Ideologically, the practical integration of the territory was also viewed as a means to integrate indigenous (and to a lesser extent Afro-Ecuadorians) into the labor market (Foote 2004). Each of these dimensions of modernity's spatiality served to enhance the performance of statehood, whose sovereignty was still not fully hegemonic through the latter half of the nineteenth century.

By the end of the nineteenth century, however, we glimpse the beginnings of an abstract space of modernity, comprising mostly the imaginative geography of what Benedict Anderson (1991) calls the "homogeneous space" of the modern nation-state. Yet sovereignty, knowledge, and everyday experiences of these spatialities were incomplete, highly uneven, and even undermined by foreign companies and neighboring states. When foreign companies began searching for oil in the 1920s, there were no accurate maps of the Amazon to work from (Foote 2004: 96). The geographical tools available were relatively rudimentary, but were established in similar ways to European countries. Ecuador seems to have been characterized, perhaps unusually, by the consolidation of geographic knowledge and skills among a small elite, and the deployment of these techniques of modernity by a defensive elite protecting their geopolitical patch.[18] Most Ecuadorian residents were excluded from these projects by racism, poverty, and lack of education.

Modernization of State Geographies: 1930s and 1940s

In these decades, geography became a tool not only for inventorizing the resources and peoples of modern nation-states, but also, as argued by David Hoosen "a necessary tool for clarifying and fostering . . . national identity (1994: 4). In Ecuador, this occurred under an increasingly centralized and professional military leadership, although civilian geographers, notably Francisco Terán, began to provide materials that nurtured a (national) geographical imagination among ordinary citizens.

In the 1920s, the military—specifically the army—had argued successfully that "knowledge of the patriotic frontiers" and an inventory of the nation's wealth (geology, hydrology, forestry, agriculture, minerals) were invaluable for the modern state. The German mission of 1925 and ongoing exchanges with Germany was followed by the 1928 inauguration of training programs in topography and cartography, and the placing of the Geographical Military Service (later to become the Geographical Military Institute, or IGM) under the army high command. The *Servicio*, later *Instituto*, had extensive links across Latin America to similar organizations, via the Instituto Pan-Americano de Geografía e Historia and the Inter-American Geodesic Service, both of which were to gain increasing significance in training and resource distribution during the Cold War.

If modernity is not just about projects and discourses, then the experience of geography as a component of people's envisioning of themselves and others began to take form in the 1930s and into the 1940s.[19] Secondary school teachers began to be taught geography: presumably for them to transmit their knowledge onto their pupils! Terán's textbook, *Geografía de Ecuador*, was first published in 1948 and went through 20 impressions, and Terán himself was appointed to the National Cultural Council in the mid-1970s. The geopolitics of these transformations was not far from the surface of mid-century Ecuadorian modernity when the 1941 war with Peru resulted in the loss of considerable territory. This event resulted in an educational fixation on what was termed "the History of the Borders," in which schoolchildren were taught that territorial reconfigurations were benign when involving Brazil and Colombia, but a visceral attack when originating in Peru. Despite increasing geopolitical and military control over state territory and geographical tools, geographical knowledges and nationalist interpretations of maps slowly came into the public arena through schools and the distribution of the "logo map" showing Ecuador's new territory (Radcliffe 1996).

Attempting Social Integration of Territory: 1960s and 1970s

During this period, we see a significant shift in the geographical imaginations and techniques of the Ecuadorian state, as the uncertain status of the country as a modern developed country came to the fore and shaped the deployment of evolving geographical knowledges and tools. However, the administration of cartographic information and map-making skills was increasingly centralized and regulated in the hands of the military. Under new legislation, the IGM held a monopoly on geographic techniques in the state.[20] The National Security Doctrine, as found in other Latin American countries, justified the creation of a higher professional training institute for geographical skills, surveillance techniques, and new geopolitical imaginations (see Hepple 1992).

Geographical techniques and knowledges were utilized in order to bring about development, which was defined in a broader, relatively inclusive way at this time, and which arguably provided the primary framework for the modern project. With oil revenues to spend, Ecuador also went through a rapid period of territorial integration, which echoed the mid-nineteenth century in some ways. Road mileage trebled between 1959 and 1978 (Quintero and Silva 1991: 238).

Ecuador's long-standing regionalism became perceived as a development "problem": in the words of an Ecuadorian geographer, "geography can help enormously to overcome many problems, principally the problems of localism. Also regionalism—for example, we have regionalism around Guayaquil and Quito" (research interview, April 1994). Much organizational, representational, and spatial imagining work went on to attempt to overcome such regionalism. Benavides (2004) shows how practices and representations of the Cochasqui archaeological site in Ecuador attempt to tie in the pre-Colombian territory to the modern nation and its citizens.

In summary, geographical practices—cartography, inventories of landscapes, territorial definition, encouraging specific imaginative geographies—are constitutive of the nature of modern state power. In the case of Ecuador, modernity was a spatial project, entailing place-specific discourses and the embodiment of modernity's identities of development and nationalism. If this discussion has been at a remove from questions about political economy and the colonial modern, it has been so in order to demonstrate the limits of totalising accounts associated with Marxism or postcolonialism. Yet the geographies of political economy explain the distribution of railway and road networks, just as colonial history informs the discourses, racializations, and gendering of modernity in the country.

Development

Choosing between . . . modernisation or local traditionalism is an untenable simplification.

(Calderón 1995: 45)

Development can be seen as a particular moment in modernity defined in classic terms that are Eurocentric and diffusionist. In the 1950s, mainstream development thinking around modernization viewed European and American development as "an independent logic and dynamism" (Slater 2004: 11) extending inexorably and unchanged across the global South. My final section thus brings the discussion of modernity's spatiality alongside recent debates about development in

Latin America. Development is used here in the sense of long-run secular improvements in economic and social well-being indicators for the majority of populations and discourses about the desirability of certain directions of change. For W. W. Rostow, modernization theorist and anticommunist author of *Stages of Growth*, development as modernization would replace colonialism (Slater 2004: 62). Development thereby reworked the meanings around tradition and the modern, viewing certain topographies or population groups as obstructive of the process of market integration and rational education/professionalism (e.g., Orlove 1993).

Yet an understanding of development as a complex spatiality of connection, shifting discourses, and experiences of constantly reconnected topography is not at the forefront of debates of Latin American development when we look at the post-development accounts. Post-development writers (sometimes called antidevelopment) reject mainstream development as an all-encompassing (Western) and undifferentiated domination representing the dark side of modernity (Rahnema with Bawtree 1997). Post-development has generated considerable debate in development studies and beyond, and is associated with various Latin American studies writers. Arturo Escobar on Colombian development projects and the Peruvian NGO PRATEC are just two examples of a broad subgroup of development critics (Escobar 1995; Apffel-Marglin with PRATEC 1998). As a pro-traditionalist reaction against modernity, post-development writing generally tends to deny the flexibility and incommensurability of different rapidly evolving development paradigms with shifting criteria of what counts as "modern" and as "development" (compare Radcliffe 2005b). Although proclaiming the virtues of (local) cultural difference, post-development approaches tend to reduce it to an account of anti-Western politics, thereby ignoring hierarchies within local cultures and the nondevelopment geographies of connection that shape "local" lives (e.g., Escobar 2001; compare Andolina et al. 2005). Post-development's simplification of development's geographies as a unilinear boundary between a singular Western culture of modern development versus local authenticity downplays both the role of nationalism in development programs (Gupta 1998), and development's actions in a global arena.[21] As with diverse *indigenista* or Catholic writers, post-development accounts see Latin American/Andean culture as everything that development is not (compare Larraín 2000). Latin America has long had an ambiguous attitude toward modernity's project of development. Long fluctuating between a desire for modernity—often in the European mode—and rejecting it as too European and unsuited for regional realities, Latin American intellectuals have often contradictorily rejected European domination while internalizing its civilizing mission (Coronil 1997: 73). As Jorge Larraín points out (2000), many Latin American writers have viewed modernity as fundamentally European and hence alien to the region. In light of this intellectual tradition, Arturo Escobar's writings might be read less as a post-structuralist critique of hegemonic development than as a regionally positioned rereading of modernity's spatiality and power.

An account of the contingent and highly uneven nature of development can acknowledge the reversals and nonlinear nature of multiple modernities, even as it draws upon an understanding of development as a set of achieved projects (falling death and morbidity rates; increased education and so on). In contrast to post-development accounts of Latin America's subjection to a singular modern development project, an account of geographical practices and spaces shows how the subjects and spaces of modernity have shifted profoundly even in the latter half of the twentieth century. For example, top-down measures of modernization planning and growth poles offered widely different experiences and projects to colonization programs into Amazon areas, or regional development plans. Again, the shift toward decentralization in the 1990s—although framed by the global geopolitics of neoliberal macroeconomic policy—has democratized access to geographical techniques such as GIS and remote sensing, as exemplified by indigenous control of GIS data in one Ecuadorian development project.

However, given the dynamic constant rearrangement of political economies and cultural politics, development has no clear guaranteed trajectory toward modernity. Although mid-twentieth-century expectations were that development would continue inexorably once set in train, places around the world demonstrate that reversals in development can occur (Ferguson 1999). Latin America's "lost decade" of the 1980s illustrates this starkly. Between 1945 and 1970, economic growth and consolidation of democracies in many areas of the region seemed to indicate that modernity had arrived. Yet in the 1980s social development was reversed and military (authoritarian) governments replaced democracies, calling into question any simple projection into modernity.

Conclusions

The spaces of modernity are fundamental to the making of modernity. As not-to-be-taken-for-granted achievements, the spaces of modernity and geographical practices lie at the center of modernity's projects, are the subject of numerous contested discourses, and underpin the embodied experiences of modernity. Where and when Latin America was modern can be answered only by reference to the spatiality of modernity and the geographical practices of diverse parts of the region. Latin America was first modern as it was positioned in the specific stretched-out geography between the Islamic modernity of the fifteenth century, the Iberian Peninsula, and the imaginative geographies of El Dorado. Although modernity has often been defined on a global canvass—the West versus the Rest—this chapter has attempted to show that modernity is about spaces and processes at a number of interconnected scales. The uneven development of capitalism began even in the colonial period to differentiate among subregions of Latin America, and actively reworked very uneven landscapes. Modernity was thus *never* going to be a homogeneous surface of development over the region's countries. Having explored the spatiality of modernity and geographical practices, we can also argue that the operation of modernity's power lies precisely in the existence of comparisons between modern and non-modern spaces at a number of scales, from the human scale to the global. Never homogeneous over space, modernity constantly uses the languages of comparison within its complex spatiality. Although in the discipline of geography there has been a tendency to explain this in terms of uneven capitalist development (e.g., Harvey 1989), this chapter has argued that it is possible to view this comparative narrative—a view from somewhere—as a cultural, social, and historical product by drawing on recent accounts of multiple reworked spaces of modernity.

Modernity is a geography, a spatial story about comparisons between different places and landscapes. Moreover, as modernity provides the topographic infrastructure for comparisons and projects for change—projects enacted in order to move toward modern status—so too the contours of modernity are constantly shifting. The reference points—or to extend my metaphors further, the triangulation points—by which modernity can be mapped are located in places that are variously reworking, aspiring to, or rejecting (certain visions of) modernity. In other words, rather than a static two-dimensional map of modernity—epitomized perhaps by the cartographic conventions used for *per capita* income measures—we are faced with a constantly mobile four-dimensional surface. In such a surface, it is the connections between places—bodies, cities, countries—and their negotiation of modernity that define the content of modernity, providing the concept with meaning.

Notes

* Many thanks to Nicola and Stephen for their invitation to the conference at which this essay was presented. In preparing the conference paper, I benefited enormously from suggestions, readings and comments from Liz Drayson, Phil Howell, and Heidi Scott, and my final-year students kept me focused on explanation. In revising this chapter, I am grateful to Luciana Martins, Laurence Whitehead, Steve Legg, and Into Goldschmidt for helpful points and for broadening my understanding of the issues.

1. In development geography, see, e.g., Pred and Watts (1992); Popke and Ballard (2004); Wright (2003); Gidwani and Sivaramakrishan (2003), and Korff (2001). In urban geography, see, e.g., Ogborn (1998), Gilbert et al. (2003); and Taylor (2000).

2. I owe this tripartite understanding of modernity's complexity to Phil Howell, who nonetheless is not responsible for the way in which I have developed it in relation to Latin America.

3. Enlightenment ideas of liberty, equality, and scientific positivism were espoused vigorously in eighteenth and nineteenth century Latin America, with intellectuals and political leaders struggling for political and cultural modernity and the reconstitution of identity. Values from the French Enlightenment, British liberalism, and Comte's positivism all crossed the Atlantic and informed intellectual development and political action (Hale 1996). In the adoption and discussion of Enlightenment ideas, Quijano (1995) sees no difference between Europe and Latin America at this stage. Yet modernity discourses in Latin America have, since the early nineteenth century, had to respond to discourses around identity, Catholicism, and the regional specifics that make European modernity inappropriate for the region. Doubts began to be raised then and throughout the twentieth century about the appropriateness of subordinating regional identity to the goals/practices of modernity (Larraín 2000).

4. *Ukhupacha* means interior world, and *runa* means people, so this phrase can be glossed as inward/religious-oriented people.

5. Spatiality basically refers to the notion, mentioned earlier, that space and society are coproduced, that society does not occur on the head of a pin and that diverse economies and societies produce very different types of spatial order, inscribed with relations of power-knowledge.

6. See Lefebvre (1991). Ogborn (1998) focuses on Magdalen Hospital, the street, Vauxhall Gardens, the Excise, and the Universal Register Office as exemplary of London's eighteenth-century spaces of modernity.

7. In a similar vein, Chatterjee (1993) argues the global project of modernity "claims for itself a singular universality, rationality and morality that depend on the subordination, exclusion or destruction of alternative forms of sociality, rationality and values."

8. Said's *Orientalism* (1978) describes how colonial descriptions of Middle Eastern peoples drew on oppositional categories to those used by Europeans to describe themselves.

9. In the words of Derek Gregory, how the West tells itself "stories of self-production" (2004: 4).

10. For example, Peru has replaced Spain as the Other in Ecuadorian nationalist imaginative geographies.

11. Early Iberian moves toward more systematic forms of knowledge and power had occurred on the periphery of an Islamic modernity, with its flourishing medical and engineering knowledge and a highly sophisticated literary and aesthetic culture in North Africa and the Middle East. Iberian peninsular culture defined itself in opposition to this. However, historians and Hispanists remind us too that Spain's Catholic systematization of knowledge and power had already begun in the thirteenth-century remnants of Iberian monarchies when the Spanish vernacular was systematized (the first Spanish dictionary came out in 1492) and a sense of national purpose/autonomy established. Nevertheless following overseas expansion and the process of unification initiated by the Catholic Monarchs, Spain began to acquire political, social, and economic structures that were undoubtedly "modern" in character over the course of the sixteenth and seventeenth centuries, albeit limited by the isolation incurred by the Catholic Inquisition.

12. Of course itineraries of connection are more complicated than this global reorientation would suggest. Andean indigenous people make complex transnational networks with Canadian first nations and Sammi people in Finland in order to pursue their trajectories of modernity.

13. Hence, this is not an argument about the functional integration of the nation-state for economic or political purposes.

14. The material here draws extensively on Radcliffe (2001) with the addition of more recent material.

15. For a recent historical discussion of Mexican uses of cartography and mapping, see Craib (2004).

16. Wolf, e.g., supported the Ibarra-San Lorenzo railway project, seeing Esmeraldas province as a rival in natural resources to the Amazon (Foote 2004: 91).

17. Later presidents, including Placido Caamaño and Eloy Alfaro, were similarly enthusiastic about the railway, Alfaro insisting "a nation without railways, highways or paved roads is a country dead to progress" (quoted in Foote 2004: 88). The ongoing fixation on the railway perhaps reflected the fact that, unlike most Latin American countries, Ecuador retained a non-coastal capital city despite the new maritime based economic and political interconnections opened up by independence from Spain (cf. Robinson 1989: 169).

18. See Silva (1995) on Ecuadorian narratives of territorial ungovernability under white-creole control.

19. A related aspect was the proposed creation of the Office of Social-Biological Statistics in 1941, envisaged as a way to inventorize regions and their inhabitants in advance of a census (Foote 2004: 180).

20. One consequence of Cold War security concerns was that the sale of maps without the Rio Protocol Line (and hence the pre-1941 territory) was illegal, punishable with up to 16 years in prison.

21. Rather than view development as a Northern imposition in a homogeneous South, Morag Bell calls for a project "tracing the spatial genealogy of ideas [which] is central to the postcolonial critique" (Bell 2002: 77), highlighting the contingency of North–South relations; shifting contexts and contents of Northern views; the varied sites and agents of knowledge; and the complex utility of development to North and South.

References

Agnew, John (1987). *Place and Politics: The Geographical Mediation of State and Society.* Boston: Allen and Unwin.

—— (1999). "The New Geopolitics of Power," in D. Massey, J. Allen, and P. Sarre (eds.), *Human Geography Today.* Cambridge: Polity.

Anderson, Benedict (1991). *Imagined Communities.* London: Verso.

Andolina, Robert, Radcliffe, Sarah, and Laurie, Nina (2005). "Development and Culture: Transnational Identity Making in Latin America." *Political Geography* 24(6) (August): 678–702.

Apffel-Marglin, Frédérique with PRATEC (1998). *The Spirit of Regeneration: Andean Culture Confronting Western Notions of Development.* London: Zed.

Bauman, Z. (1991). *Modernity and Ambivalence.* Ithaca: Cornell University Press.

—— (1999). *In Search of Politics.* Cambridge: Polity.

Bell, Morag (2002). "Inquiring Minds and Postcolonial Devices: Examining Poverty at a Distance," *Annals of the Association of American Geographers* 92 (3): 507–523.

Benavides, O. Hugo (2004). *Making Ecuadorian Histories.* Austin: University of Texas Press.

Bonnett, A. (2002). "The Metropolis and White Modernity." *Ethnicities* 2(3): 349–366.

Calderón, Fernando (1995). "Latin American Identity and Mixed Temporalities: or How to be Post-Modern and Indian at the Same Time," in J. Beverley, J. Oviedo, and M. Aronna (eds.) *The Postmodernism Debate in Latin America.* London: Duke University Press, pp. 55–64.

Chatterjee, Partha (1993). *The Nation and Its Fragments: Colonial and Postcolonial Histories.* London: Princeton University Press.

Coronil, Fernando (1997). *The Magical State.* Chicago: University of Chicago Press.

Craib, Raymond (2004). *Cartographic Mexico: A History of State Fixations and Fugitive Landscapes.* London: Duke University Press.

Dalby, Simon (2003). "Calling 911: Geopolitics, Security and America's New War." *Geopolitics* 8(2): 61–86.

Davis, Diane (1999). "The Power of Distance: Re-Theorizing Social Movements in Latin America." *Theory and Society* 28: 585–638.

De la Fuente, Rosa (2004). *La autonomía indígena en Chiapas: La construcción de un nuevo espacio de representación.* Unpublished Ph.D. thesis, Universidad Complutense de Madrid, Department of Political Science, Madrid.

Dore, Elisabeth and Molyneux, Maxine (eds.) (1999). *Hidden Histories of Gender and the State in Latin America.* London: Duke University Press.

Dussel, Enrique (1993). "Eurocentrism and Modernity." *Boundary 2* 20 (3): 65–76.

—— (2002). "World-System and 'Trans'-Modernity." *Nepantla: Views from the South* 3: 2.

Escobar, Arturo (1995). *Encountering Development: The Making and Unmaking of the Third World.* Princeton: Princeton University Press.

—— (2001). "Culture Sits in Places: Reflections on Globalism and Subaltern Strategies of Localisation." *Political Geography* 20: 139–174.

Ferguson, James (1999). *Expectations of Modernity: Myths and Meanings of Urban Life on the Zambian Copperbelt.* Berkeley: University of California Press.

Foote, Nicola (2004). *Race, Nation and Gender in Ecuador: A Comparative Study of Black and Indigenous Populations, c.1895–1944.* Unpublished Ph.D. dissertation, University of London.

Foucault, Michel (1984). "Space, Knowledge and Power," in Paul Rabinow (ed.), *The Foucault Reader.* New York: Pantheon, pp. 239–56.

García Canclini, Néstor (2000). "From National Capital to Global Capital: Urban Change in Mexico City." *Public Culture* 12(1): 207–213.

Giddens, Anthony (1990). *Consequences of Modernity.* Cambridge: Polity.

Gidwani, V. and Sivaramakrishnan, K. (2003). "Circular Migration and the Spaces of Cultural Assertion." *Annals of the Association of American Geographers* 93(1): 186–213.

Gilbert, David, Matless, David, and Short, Brian (eds.) (2003). *Geographies of British Modernities: Space and Society in the Twentieth Century.* Oxford: Blackwell.

Gregory, Derek (1994). "Modernity," in R. Johnston et al. (eds.), *The Dictionary of Human Geography.* Oxford: Blackwell.

—— (2004). *The Colonial Present.* Oxford: Blackwell.

Guano, Emanuela (2002). "Spectacles of Modernity: Transnational Imagination and Local Hegemonies in Neoliberal Buenos Aires." *Cultural Anthropology* 17(2): 181–209.

Gupta, A. (1998). *Postcolonial Developments: Agriculture in the Making of Modern India.* Oxford: Oxford University Press.

Hale, Charles (1996). "Political Ideas and Ideologies in Latin America, 1870–1930," in L. Bethell (ed.), *Ideas and Ideologies in Twentieth Century Latin America.* Cambridge: Cambridge University Press.

Hansen, Thomas B. and Stepputat, Finn (2001). "Introduction: States of Imagination," in T. B. Hansen and F. Stepputat (eds.), *States of*

Imagination: Ethnographic Explorations of the Postcolonial State. Durham: Duke University Press.

——— (eds.) (2005). *Sovereign Bodies: Citizens, Migrants and States in a Postcolonial World.* London: Princeton University Press.

Hardt, M. and Negri, T. (2000). *Empire.* London: Harvard University Press.

Harley, J. B. (1992). "Rereading the Maps of the Columbian Encounter." *Annals of the American Association of Geographers* 82(3): 543–65.

Harvey, David (1989) *The Condition of Postmodernity.* Oxford: Blackwell.

Hepple, Leslie (1992). "Metaphor, Geopolitical Discourse and the Military in South America," in J. Duncan and T. Barnes (eds.), *Writing Worlds.* London: Routledge.

Hoosen, David (ed.) (1994). *Geography and National Identity.* Oxford: Blackwell.

Howell, Philip (forthcoming). *Prostitution, Empire and Modernity: The Geography of Regulation, 1850–1930.* Cambridge: Cambridge University Press.

Jervis, John (1998). *Exploring the Modern: Patterns of Western Culture and Civilisation.* Oxford: Blackwell.

Joseph, Gilbert and Nugent, Daniel (eds.) (1994). *Everyday Forms of State Formation: Revolution and the Negotiation of Rule in Modern Mexico.* London: Duke University Press.

Korff, R. (2001). "Globalisation and Communal Identities in the Plural Society of Malaysia." *Singapore Journal of Tropical Geography* 22(3): 270–283.

Larraín, Jorge (2000). *Identity and Modernity in Latin America.* Cambridge: Polity.

Lechner, Norbert (1995). "A Disenchantment called Post-Modernism," in J. Beverley, J. Oviedo, and M. Aronna (eds.), *The Postmodernism Debate in Latin America.* London: Duke University Press, pp. 147–64.

Lefebvre, Henri (1991). *The Production of Space.* Oxford: Blackwell.

Mahoney, James (2003). "Long-Run Development and the Legacy of Colonialism in Latin America." *American Journal of Sociology* 109(1): 50–106.

Marston, Sallie (2000). "The Social Construction of Scale." *Progress in Human Geography* 24(2): 219–242.

Martins, Luciana and Abreu, Mauricio (2001) "Paradoxes of Modernity: Imperial Rio de Janeiro, 1808–1821." *Geoforum* 32: 533–550.

Mignolo, Walter (2000). *Local Histories/Global Designs: Coloniality, Subaltern Knowledges and Border Thinking.* London: Princeton University Press.

Ogborn, Miles (1998). *Spaces of Modernity: London's Geographies, 1680–1780.* London: Guildford Press.

Oporto, Henry (1992). "Es posible una Bolivia moderna?" *Revista UNITAS* 7(September): 84–92.

Orlove, Ben (1991). "Reading the Maps and Mapping the Reeds: The Politics of Representation in Lake Titicaca." *American Ethnologist* 18(1): 3–38.

———— (1993). "Putting Race in Its Place." *Social Research* 60(2): 301–36.

Oslender, Ulrich (2004). "Fleshing Out the Geographies of Social Movements: Colombia's Pacific Coast Black Communities and the "Aquatic Space." *Political Geography* 23(8): 957–85.

Perreault, Tom and Martin, Patricia (2005). "Geographies of Neoliberalism in Latin America." *Environment and Planning A* 37, February.

Popke, E. and Ballard, R. (2004). "Dislocating Modernity: Identity, Space and Representations of Street Trade in Durban, South Africa." *Geoforum* 35(1): 99–110.

Power, Marcus (2003). *Postcolonial Development Geographies.* London: Longman.

Pred, Allan (1984). "Place as Historically Contingent Process: Structuration and Time-Geography of Becoming Places." *Annals of the Association of American Geographers* 74: 279–97.

Pred, Allan and Watts, Michael (1992). *Reworking Modernities: Capitalism and Symbolic Discontent.* New Brunswick: Rutgers University Press.

Quijano, Anibal (1995). "Modernity, Identity and Utopia in Latin America." in J. Beverley, J. Oviedo, and M. Aronna (eds.), *The Postmodernism Debate in Latin America.* London: Duke University Press, pp. 201–16.

Quintero, Luis and Silva, Erika (1991). *Ecuador: una nación en ciernes.* Quito: FLACSO.

Radcliffe, Sarah (1996). "Imaginative Geographies, Post-Colonialism and National Identities: Contemporary Discourses of the Nation in Ecuador." *Ecumene* 3(1): 21–42.

———— (2001). "Imagining the State as a Space: Territoriality and the Formation of the State in Ecuador," in T. Blom Hansen and F. Stepputat (eds.), *States of Imagination: Ethnographic Explorations of the Postcolonial State.* London: Duke University Press.

Radcliffe, Sarah (2005a). "Neoliberalism as We Know It, but Not Under Conditions of Its Own Choosing: A Commentary" *Environment and Planning A* 37, February.

———— (2005b). "Development and Geography II: Towards a Postcolonial Development Geography?" *Progress in Human Geography* 29(3): 291–98.

Radcliffe, Sarah and Westwood, Sallie (1996). *Remaking the Nation: Place, Politics and Identity in Latin America.* London: Routledge.

Rahnema, M. and Bawtree, V. (eds.) (1997). *The Post-Development Reader.* London: Zed.

Robinson, David (1989). "The Language and Significance of Space in Latin America," in J. Agnew and J. Duncan (eds.), *The Power of Place: Bringing Together Geographical and Sociological Imaginations.* London: Unwin Hyman, pp. 157–84.

Said, Edward (1978). *Orientalism.* New York: Vintage Books.

Scott, Heidi (2003). "Contested Territories: Arenas of Geographical Knowledge in Early Colonial Peru." *Journal of Historical Geography* 29(2): 166–88.

Silva, Erika (1995). *Los mitos de la ecuatorianidad: ensayo sobre la identidad nacional.* Quito: Abya Yala.

Slater, David (1998). "Rethinking the Spatialities of Social Movements: Questions of (B)orders, Culture and Politics in Global Times," in S. Alvarez, E. Dagnino, and A. Escobar (eds.), *Cultures of Politics, Politics of Culture: Re-Visioning Latin American Social Movements.* Boulder: Westview Press, pp. 380–401.

—— (2004). *Geopolitics and the Post-Colonial: Rethinking North-South Relations.* Oxford: Blackwell.

Smith, Neil (1984). *Uneven Development.* Oxford: Blackwell.

Stephenson, Marcia (1999). *Gender and Modernity in Bolivia.* Austin: University of Texas Press.

Stoler, Ann Laura (1995). *Race and the Education of Desire: Foucault's History of Sexuality and the Colonial Order of Things.* London: Duke University Press.

Taylor, P. J. (2000). *Modernities: A Geohistorical Interpretation.* Cambridge: Polity.

Terán, Francisco (1983). *Estudios de historia y geografía.* Quito: Biblioteca Ecuatoriana.

Wright, M. W. (2003). "Factory Daughters and Chinese Modernity: A Case from Dongguan." *Geoforum* 34(3): 291–301.

Young, Robert (1990). *White Mythologies.* London: Routledge.

Chapter 2

Modernity and Tradition: Shifting Boundaries, Shifting Contexts

Peter Wade

Definitions of Modernity

In approaching the question of the chronology of modernity in Latin America, I attempt to explore what might constitute an anthropological approach to modernity, particularly in Latin America but also more widely. Before I tackle that point directly, it is worth exploring briefly some definitions of modernity and its periodizations—which would presumably be the kinds of definitions and timescales that anthropologists would also use. Many commentators point to a lack of consensus on these matters. Smart (1990: 15) notes "the presence of a constellation of related terms, [and] a lack of specificity associated with the concepts employed, particularly in reference to their historical referents or periodisation" (see also King 1995; Therborn 1995). Noting that the term derives from the fifth-century Latin term, *modernus*, used to mark an official transition from the pagan to the Christian (and itself, I would add, derived from *modus*, measure or manner), Smart observes that the term is used thereafter "to situate the present in relation to the past of antiquity." Citing Habermas, he says that the term appears "exactly during those periods in Europe when the consciousness of a new epoch formed itself through a renewed relationship to the ancients" (Smart 1990: 17). In an important sense, it is a relational term.

Nevertheless, most commentators prefer to adhere to some periodization. Turner (1990: 6) says that modernity "arises with the spread of Western imperialism in the sixteenth century" and continues with the dominance of capitalism, the acceptance of scientific procedures, and the separation of household from economy. Hardt and Negri (2000: 70–87) trace it to the period 1200–1600, when two contradictory forces emerged: the first based on the affirmation of the powers of this, as opposed to a transcendent, world and of human subjects within it to make themselves; the second based on attempts to reestablish transcendent (although not necessarily other-worldly) control over those powers, attempts that arrived at the concept of modern sovereignty. Berman (1983) outlines three successive waves of change that affect Europe and the world from the early seventeenth century, but reserves the term modernity for the post-Enlightenment nineteenth-century period, when, as Williams (1988) notes, "modern" acquired a positive meaning. Escobar (2004: 211) says that, for dominant theories of modernity in general, "modernity has identifiable temporal and spatial origins: 17th century northern Europe, around the processes of Reformation, the Enlightenment, and the French Revolution. These processes crystallised at the end of the 18th century and became consolidated with the Industrial Revolution." Scott Lash sides with theorists such as Daniel Bell in distinguishing between modernity (or, for some, modernization) and modernism. In this scheme, modernity as an economic and technological process began in the sixteenth century, while modernism, dating from the late nineteenth century, is seen as a cultural or aesthetic reaction to the contradictions created by that process (Smart 1990: 17–9). This distinction has also been seen as particularly relevant to Latin America where, for some, economic and social modernity has lagged behind or at least been in disjuncture with cultural modernism (García Canclini 1995: 41–65; Schwarz 1992).

Anthropological Approaches to Modernity

I think many anthropologists are broadly happy with these kinds of approach and do not worry too much about the periodizations and disputes. Modernity can be generally understood as "what results from the diversified impact of capitalism on social formations across the world," as Moreiras (2001: 3) paraphrases Charles Taylor, although one might want to add something about the impact of Western scientific rationality and other formalized, calculative rationalities (in the Weberian sense). Many approaches to the "anthropology of modernity" base themselves on this kind of broad view of the subject.

In my view, common anthropological approaches to modernity take two main avenues. The first is simply to expand the purview of anthropology beyond the non-Western to include an ethnography of the West. This is unobjectionable in itself and actually has a longish history within anthropology. It could, of course, imply an unexamined division between the "traditional" non-West and the "modern" West, but I think most anthropologists would reject such easy dualisms—at least at first sight. The second avenue leads to an analysis of processes of resignification and hybridization. This is about exploring how "local," usually non-Western people, adapt objects, ideas, and symbols from global circuits of production, consumption, and knowledge, indigenizing, resignifying, appropriating, and hybridizing them in the process and perhaps generating "multiple" or "alternative" modernities (Comaroff and Comaroff 1993; Englund and Leach 2000; Inda and Rosaldo 2001; Miller 1995). Debates then center around the way both homogenization and differentiation occur at the same time and how continuity and change are to be perceived and analyzed (Robbins 2004; Sahlins 1999; Wilk 1995). Again, nowadays, many anthropologists would resist a clear division between the local and the global, as if localities were neat, bounded places that simply suffered the impact of external, global forces. However, it is worth noting a certain mutual reinforcement of the spatial and the temporal here: modernity is not only "after" tradition (although it is "before" it when it comes to reaching the future), it is "beyond" locality and acts as the "wider context" for it.

Teleology and Scale

I think parallel dualisms between traditional and modern, local and global, and subordinate (subaltern) and dominant (hegemonic) often remain at a deeper level in anthropological approaches. It may be the case that, as Dunkerley (2000: 51) says, we are currently in a " 'globalised' climate, when everything on the planet is linked to everything else without hesitation or embarrassment." But I would argue that analytic linkages continue to be made in ways that betray vestiges of dualisms. I also argue that anthropology can offer ways to undo these dualisms, and their associated effects of scaling and teleology (and ultimately Eurocentrism), as long as these ways are pursued thoroughly. The paradoxical commitment of anthropology to cultural relativism (particularism) but also cultural equality (universalism) highlights the fact that all people are coeval, that "we are all modern"—or that we have all "never been modern" to adapt Latour's phrase (Latour 1993). Of course, this insight has been part of anthropology's critique of its own intellectual heritage. Insisting that everyone is equally modern is not the same as saying that Western modernity has accomplished a full sweep; rather it is a way of contesting the Eurocentric assumption that some are "ahead" of others on a scale of progress. However, the dualistic traces of anthropology's heritage have not, I think, been fully erased in the process of critique. But one can build on the insight that underlies the critique. It is well known that modernizationist and developmentalist theories tend to be Eurocentric. They are based on the assumed superiority of Western forms of economic and political organization and the teleological idea that these emerged in Europe and have diffused, or been imposed, globally from there and that they should continue to do so. As King (1995) points out, there is a spatial as well as temporal narrative here, in that modernity has a spatial origin and movement as well as an historical one (see also, Radcliffe, this volume). I would add that there is also a scaling effect at work, in which modernity is seen as large scale or global, while its assumed opposite, tradition, is seen as smaller scale and local: modernity can thus act as a "context" for tradition; it is the "bigger picture" (Englund and Leach 2000).

Critiques of such Eurocentrism and teleology are legion. García Canclini (1995: 3–7) criticizes the inadequacies of developmentalism. Dussel (1995: 66–7) argues that European modernity constituted itself in opposition to an alterity, a periphery that was concealed and misrecognized. The first of these peripheral Others was Latin America. Escobar (2004: 217–20) outlines the work of a group of Latin American scholars, including Enrique Dussel, Walter Mignolo, and Aníbal Quijano (see also Castro-Gómez 1998). They emphasize the mutually constitutive nature of modernity and coloniality (a term that includes but goes beyond simple colonialism to encompass postcolonial forms of domination). Western modernity was predicated on the domination of the non-West, especially the Americas, the conquest of which began the modern era. Location within the underside of modernity, or better, at the "border" between modernity and coloniality (Mignolo 2000), can give rise to ethical contestations of the dominant Eurocentric discourse.

Yet it is perhaps more difficult than it appears to root out teleology and scaling effects. One can argue that Latin America has alternative forms of modernity, or an uneven modernity that included what would now be called postmodern forms before postmodernity was seen as a general condition of Western societies, or that the region has complex hybrid combinations of tradition and modernity in which the latter does not displace the former. But it is harder to displace the ideas that modernity itself emerged in Europe, that it emerged there before it emerged anywhere else, that it had uneven impacts on other parts of the world that were not, by definition, (as) modern, and that it has a (more) global scale than other social-cultural forms. These ideas retain a commonsense obviousness that is hard to unsettle, yet they all imply a basic teleology that, while it does not entail a simple goal of inevitable Westernization, still constructs an historical narrative in which Western modernity comes first, shapes the world through its diffusion, and acts as the larger context for other processes.

The premise for the workshop—"When was Latin America Modern?"—that was the basis for this book is a good example of the operation of this teleological and scalar way of thinking. Agreed, the focus on chronology and on the question of when Latin American became modern, if ever, is simply a heuristic device for addressing the more complex question of what modernity is in the Latin American context. But the device supposes a modernity that developed "elsewhere" and "before," and then "arrived" in Latin America, albeit in different ways and at different times and with doubtless contradictory and unpredictable effects. There is an underlying premise of an historical narrative led by this Western modernity that exists beyond, as well as within, Latin America.

A different way of thinking about social change as non-scalar and non-teleological involves seeing all social forms as coeval and at the same scale—or rather sees all forms of periodization or historicization and all attempts at scaling (creating figure/ground devices) as constructions, whether analytic or "popular," with political effects. The idea of modernity as temporally and spatially based in Europe is a construction with evident effects in the realms of power and knowledge. Scaling certain forms as "national" or "global" gives them greater power (see Radcliffe, this volume), as they then act as the context for "local" social forms and processes. Context can then figure in different ways in relation to the local: from being an explanation to being part of the technology of spatial power for making the local respond in certain ways.

A (flawed) parallel for a non-scalar, non-teleological way of thinking can be found in understandings of biological change through natural selection. There is no direction or teleology in the so-called evolution of life processes: new forms (of organisms, DNA, population structures) emerge, but without direction. Forms may spread from one place to another, some forms spread faster than others, some are more ubiquitous than other (flies, humans, some bacteria), but this does not obey a logic of center and periphery or of temporal unilinearity. (The notion of the "anatomically modern human" is,

I would argue, a teleological term that is used in evolutionary theory despite its inconsistency with the basic principles of natural selection.) There is also no scale of forms. Some forms are bigger than others, in terms of individual phenotype; some are more numerous as populations. But each form is equally the context for all other forms; no form is "the global" to another form's "local"; all forms are equally global and local at the same time, even if some are much more ubiquitous than others. They are all elements in the same network. This is, of course, a highly complex network and it is interesting that recent work in social sciences and biological and physical sciences have found some common ground in the idea of complex networks that have properties of nonlinearity and emergent self-organization, which emerges endogenously, that is, not as a result of stimulus from the external environment (Escobar 2004: 222; Thompson 2004). Social science work tends to see such complex networks as the recent product of globalization, but one could proceed from the assumption that social forms have always been networked in this way, but they have been constantly subjected to hierarchical orderings by human agents who scale these networks and give them teleological meaning.

This parallel is (deeply) flawed insofar as the key difference between these biological processes and sociocultural ones is that humans have self-conscious agency and impose design on processes they seek to control. (Of course, notions of design and agency have crept into—and are the subject of debate in—theories about natural selection via metaphors such as the selfish gene and a focus on the individual organism construed as a maximizing reproducer of offspring.) People may *design* processes to have global and/or local impact and to be progressive and/or traditional. But the key point is that these processes are *construed* as global/local and modern/traditional by the people who create and enact them and who feel their effects. Being ubiquitous is not the same as being "global": processes such as eating and defecation are very ubiquitous but do not normally get labeled as global or globalizing. Being new is not the same as being modern: new things emerge continuously—things that are not the same as existing things (e.g., babies, conversations)—but not all these are labeled as "modern." Various scholars, from Edward Said to Mignolo and Moreiras, have pointed out that practices of academic knowledge are themselves important examples of how scales and teleologies are constructed so as to create such entities as "Europe" and "Latin America" and to place them in certain relationships of power and knowing (Castro-Gómez 1998; Mignolo 2000; Moreiras 2001).

The Persistence of Teleology and Scale

To pursue the question of underlying temporal teleology and spatial scaling persisting in examinations of modernity in Latin America, let us examine two commentaries on Néstor García Canclini's *Hybrid Cultures*. Both commentators recognize García Canclini's disavowal of modernizationist teleologies, yet both also register doubts about the completeness of this. Beverley (1999: 127) notes that for García Canclini, as for Lyotard, "there is no 'outside' of globalization from which to construct an opposition to it—no 'tradition,' no 'third world,' no 'nature,' no autonomous sphere of popular culture, no modernist hermeneutics of depth." This comment is aimed at a specific target: the idea that there is no authentic site, outside or below global capitalism, which will serve as the basis for constructing resistance. Everything is pervaded by global capitalism and resistance has to be constructed from within this space. But the comment also accords nicely with my argument that we need to think about social change in a way that is non-scalar and non-teleological; the world is not separated into global and local; tradition (and modernity) are ways of reading or construing processes of change, not just things that exist out there. (I read the idea of a "modernist hermeneutics of depth" as akin to a notion of scale.) In his Foreword to *Hybrid Cultures*, Renato Rosaldo is more skeptical about García Canclini's approach:

> Much work in Latin America employs the distinction between the modern and the traditional more as an organising assumption than as a topic for investigation. Yet the distinction is as vexed to me as it is clear to García Canclini. Not unlike notions of the global and the national or the modern and the postmodern, it is evident that both social forces operate in the present and that both are empirically difficult to separate. [. . .] When García Canclini argues that the processes of production and consumption imply that no realm of cultural production can remain independent of the marketplace (and vice versa), it should follow that entering and leaving modernity deconstructs—indeed, dissolves into hybridity—the very distinction between tradition and modernity that he resolutely maintains. (Rosaldo, in García Canclini 1995: xv)

Rosaldo thinks that, although García Canclini traces very well the way the modern and the traditional hybridize in Latin American contexts, the concepts of modernity and tradition themselves remain more or less unscathed. My own reading of *Hybrid Cultures* is less clear on this issue. Certainly, the concepts of tradition and modernity are often deployed in a way that takes for granted what each term means and what it encompasses. On the other hand, parts of the analysis also destabilize both concepts—for example, in the critique of "tradition" as deployed in folklore studies (García Canclini 1995: 147–70).

Beverley mounts a slightly different critique—one that could be read as contradicting his comments cited earlier. In *Hybrid Cultures*, he detects the operation of a "postnational teleology," which is not dissimilar to the national teleology that has been identified in ideologies of *mestizaje* (Beverley 1999: 127): "Hybridisation functions . . . as a process of dialectical sublation or transcendence of prior states of dissonance or contradiction in the configuration of a subject, social group or class, national or regional identity. In this sense, Canclini's [*sic*] argument is essentially modernist, rather than anti- or postmodernist, as it appears at first sight" (127). Again, I am not sure this critique sticks entirely. García Canclini is, in my view, equivocal about whether processes of hybridization reinforce existing power inequalities or help to ameliorate them (García Canclini 1995: 227–8, 239, 241): there is no necessary transcendence or sublation.

If the critiques of Beverley and Rosaldo are only partially on target, they nevertheless serve to show that dualisms, teleology and scaling effects may linger—or be seen to linger—in recent anthropological approaches to modernity. Seeing Latin American cultures as complex hybrids of modernity and tradition tends to assume that we know what constitutes tradition and modernity in the first place, in order to be able to think the possibility of the hybrid. Tradition ends up being the local, what was there "before" modernity arrived; modernity ends up being construed as something "out there" in the global world, which arrives on the scene. But what if the "traditional" is itself a product of global networks that have been operating in the very long term? What if the "modern" is itself constituted through these same networks and thus not easily distinguishable in temporal and spatial terms from the traditional? Then our attention is turned to how and why the concepts of modernity and tradition are being constructed and deployed, in both academic and nonacademic thinking. As García Canclini says (1995: 141), "All culture is the result of a selection and a combination—constantly renewed—of its sources. In other words it is a product of a staging."

Englund and Leach give further examples of the persistence of hidden "meta-narratives of modernity" in anthropological work on "multiple modernities." Their empirical material comes from Africa and Melanesia, but their argument is relevant to other regions too.

Three assumptions, they say, underlie work on multiple modernities:

> The first is that modernity, full-fledged and recognizable, is everywhere. This assumption precludes teleology; some parts of the world are not somehow less modern than others. The second is that the institutional configuration of modernity cannot be defined in advance. The analyst may choose to highlight witchcraft in one setting, aesthetics in another, and political economy in a third . . . The third is that diverse cultures persist, offering, according to some perspectives, "local" responses to "global" processes.

However, they argue, this approach "cannot obliterate the logical requirement of representing variation against something that is *invariable*" (Englund and Leach 2000: 228, emphasis in the original).

This invariably boils down to a meta-narrative of rupture in which the "wider context" of capitalism impinges on local peoples and produces certain reactions, which are intelligible (especially to the analyst) from the privileged perspective of that context. Thus, for example, it is possible to see Melanesian people's concern with money, organ theft, death, and white people as a local reaction, through local cultural lenses, to the encroachment of capitalism—there are obvious parallels with Taussig's argument about the devil and commodity fetishism (Taussig 1980). Leach argues that their concerns can better be explained in terms of their own cultural understandings of productivity, relationships, and bodies; the "wider context" is not always that of global capitalism.

The Constructedness of "Tradition" and "Modernity"

Anthropology gives various examples of the importance of these processes of construction. Kuper's book, *The Invention of Primitive Society: The Transformation of an Illusion* (1988), traces how anthropology consistently reinvented the notion of "primitive society" as a basis for its intellectual endeavors. It needed the idea of some original state of social being, even when social evolutionary paradigms had been discredited in the discipline. Fabian's *Time and the Other: How Anthropology Makes Its Object* (1983) is a more radical critique of what he sees as the Orientalist project underlying much of anthropology's history, which located the primitive, the traditional, and the Other in the past, even when they were being studied in the present.

A specific example is that of debates about hunter-gatherers. These debates have not focused on Latin America—although they could have done (Pagden 1982)—but I think the digression is worthwhile.

The academic debate focused on the San (also known as "Bushmen") of the Kalahari desert in southern Africa. One set of anthropologists, the "traditionalists," were accused by the "revisionists" (who were of a Marxist bent), of creating the figure of the hunter-gather tribe, which, according to the traditionalists, still persisted today in parts of the Kalahari and could, with care, be used to make reasonable conjectures about the hunting-gathering way of life as an ecological adaptation that has characterized humans for most of their evolutionary history. Of course, the traditionalists recognized that modernity had in the last few decades taken a huge toll on these relatively isolated peoples. The revisionists argued that, despite appearances to the contrary, the hunter-gatherers had for centuries been integrated into regional and global processes of change, including capitalism, and that their present condition of apparent isolation and "traditionality" was actually a product of this integration, which had, as it were, chewed them up and spat them out, having sucked the desert dry of its useful resources (Kuper 1993; Solway and Lee 1990; Wilmsen and Denbow 1990).

The debate was acrimonious and polarizing, but it served to show how the figure of the hunter-gatherer could function as an academic construction with which to make arguments about the powers of modernity: either modernity (specifically in the shape of capitalism) had limited powers, which had been braked by the desert fastness of the Kalahari and the determined autonomy of the San; or it was omnipotent, integrating everything and converting the San into desert proletarians. In a review of the debate, Stiles (1992) concluded that hunter-gather peoples all over the world had probably been integrated in uneven ways into much larger circuits of exchange, production, and consumption for some 2,000 years. Rosaldo (1982) argues that so-called primitive peoples may be used as a figure to think about humanity and especially about "modern" people. Questions about "human nature"—violence/peacefulness, territoriality/sharing, patriarchy/gender equality—have often been debated with the figure of the "primitive hunter-gatherer" representing all that is supposedly natural. Stiles's even-handed approach, which assumes that situation normal is one of very long-term codependence, and (uneven) integration into global circuits of exchange, strikes me as a refreshing challenge to common assumptions both about the isolated hunter-gatherer and about the insidiously transformative omnipotence of capitalism.

In the Latin American context, Taussig (1987) has argued that the figure of the primitive or wild Amazonian native has been constructed and deployed—with the participation of some Amazonian people—as an alter against which to think about other forms of indigeneity (e.g., Andean) and colonial and postcolonial forms of "civilization." His argument is more about the discursive construction of images of Amazonian indigeneity and less about the way Amazonian society has itself been shaped in long-term interactions and integration, although this is implicit in his references to long-term trade in magic and healing. Murphy's study of the Mundurucú, a group living in the lower Amazon valley in Brazil, is more telling in this respect, showing how their social organization, as apparent to anthropologists in the 1950s, was actually the result of changes that had taken place since the eighteenth century, involving their interactions with Spanish colonists (see Wolf 1982). What appeared to be a local tradition was the product of integration into global networks.

In my own work on the history of Colombian popular music in the twentieth century, I was struck by the concern with "tradition" shown by all those involved with music, whether as composers, players, DJs, record industry personnel, academics, or listeners and dancers. It was very common to assert that the popular music styles that had emerged at various times were a mixture of "traditions" (typically, African, indigenous, and European) and especially of the "traditional" with the "modern." For example, a given style of music was construed by observers as "traditional" to the context of the Caribbean coastal region of Colombia in the mid-nineteenth century. As such it could serve as an authentic basis on which to create more modern hybrids— which then themselves served as the basis for yet further hybrids in the twentieth century. Such was the narrative about *porro*, the history of which was typically narrated as an indigenous local tradition, hybridized in mid- and late nineteenth century into a brass-band style, which was then hybridized into a jazz-band style in the mid-twentieth century (and later subjected to re-traditionalization in the staging of local traditions of folklore). I argued that the mid-nineteenth-century forms were themselves formed in complex interactions involving Caribbean and European musical practices (Wade 2000). The "traditional" was as modern as the "modernity" with which it was hybridizing. Each hybrid was seen post hoc as traditional.

Foregrounding/Backgrounding

In this constructionist approach—doubtless all too predictable from a social anthropologist—I am not trying to argue that tradition and modernity are "merely" discursive constructions. Capitalism, science, secular critique, and so on are real social processes that have a history and geography. One can legitimately pose the question of how these processes have occurred in Latin America. I am arguing against a particular way of conceptualizing the processes in question, a way that reproduces subtle dualisms, that maintains a hidden teleology and scaling effect, and that glosses over the manner in which "tradition" and "modernity" are both analytic and popular constructions that foreground and background certain things.

I assume an anthropological approach to these issues to be one that asks what difference it makes to our analysis if we see tradition/modernity as a discursive analytic couplet. What is hidden and made visible when that couplet is constituted in different ways? If we try to avoid teleology and scaling effects, what might our analysis look like?

Take the example of the idea of race in Latin America. It is common to analyze this question in terms of how ideologies and scientific knowledge about race, coming from Europe and the United States in the nineteenth century and into the twentieth, arrived in and shaped Latin American intellectuals (and others), perhaps in the process being adapted to suit the particular context of the region in general or a nation in particular (Graham 1990; Stepan 1991). Also common are reflections on how developments in U.S. race relations have shaped Brazilian ideas about race, whether contrastively (we Brazilians are not racist like them) or by inspiration (we black Brazilians need to be racially self-conscious like African Americans) (Fontaine 1981; Winant 1992).

This is all valuable stuff, but one could also think about how racial science in the United States and Europe was itself shaped by what was going on in Latin America. Obviously, European thinkers were developing their ideas with a global vision of the "races" of the world, but I am thinking more of the way ideas about race in Latin America fed into European ideas, or rather how different ideas were being constituted in a transnational dialogue, which blurred easy dualisms between modern and non-modern, or between local and global— even if the scientists concerned in Latin America and Europe had clear views about where modernity and progress lay. Dunkerley's resolvedly multilateral and chronology-challenging *Americana* surely moves in this direction (Dunkerley 2000). Poole also attempts a history of race and vision that undermines a conventional teleology of modernity: "The currents determining what 'modernity' would be . . . did not flow only in one direction. Rather, the sentiments, practices, and discourses known as European modernity were themselves shaped in important ways by the constant flow of ideas, images and people between Europe and the non-European—in this case Andean— world" (1997: 21). In relation to race in particular, she argues that ideas about racial difference did not simply persist from colonial times through to modern times (which, in her Foucauldian timeline, began in the late eighteenth century). Instead, race became visible and seen as natural in new ways in this multilateral constitution of modernity. Images of the Andes and Andean people were important in this process, influencing key thinkers such as Buffon and Humboldt, and feeding both European and Latin American thought on racial difference with hundreds of photographic images of Andean "racial types," taken by both European and Latin American photographers, circulating through Europe and the Americas (Poole 1997: Chs. 3, 5).

Micol Siegel's study of the mutual, transnational constitution of ideas about race in Brazil and the United States in the twentieth century is another excellent example of what I have in mind (Siegel 2001). She shows how U.S. ideas about race were being formed in relation to Brazil, including by means of visits of black and white U.S. intellectuals to Brazil (despite covert restrictions on the travel of U.S. blacks to the country). At the same time, the reverse process was occurring. U.S. black intellectuals sometimes used the image of racial democracy in Brazil (which they thought their visits reinforced for them) as a means to critique U.S. racial policies. Brazilian black intellectuals tended to avoid more assertively racial stances and to buy into national ideologies of democracy, reinforcing these with liberal ideals about equality and freedom.

This kind of approach shares much with Gilroy's well-known thesis that black identity was constructed in diasporic circuits of exchange across the Atlantic over a period of 150 years.[1] More than this, Gilroy's work argues that modernity itself was constituted in part through these black Atlantic exchanges, for example, through the critiques that black intellectuals made of a Western modernity based on slavery and racism. Social locations often construed as "traditional" and "local," such as black communities in the United States and Caribbean, were actually constitutive of the processes of modernity, such as "double consciousness": "It is being suggested that the concentrated intensity of the slave experience is something that marks out blacks as the first truly modern people, handling the nineteenth century dilemmas and difficulties which would become the substance of everyday life in Europe a century later" (Gilroy 1993: 220–1).

Matory (1999) criticizes Gilroy for neglecting the African side of the equation. Matory shows how Brazilian Candomblé religious centers (which are often seen as, or claim to be, "pure African") were shaped very strongly by Brazilian blacks who went to Africa, where they were educated in English-speaking Presbyterian schools, often became Freemasons, and also made visits to England. They were proud of their English connections and sometimes adopted Anglicized names. In Nigeria, they imbibed Yoruba culture and religion and transmitted that interest and knowledge on to Candomblé practices in Bahia. In turn, the whole position of Yoruba culture and religion in Lagos in the mid-nineteenth century was shaped by returnee migrants from Brazil (some expelled from Brazil as rebels), who found some shelter in the British protectorate in Lagos (and were joined there by free blacks from Sierra Leone, the United States, and Jamaica). These migrants helped local African intellectuals and cultural activists create the Yoruba nation as a self-conscious entity. If black Atlantic culture was constitutive of modernity, then that culture was also being made in complex exchanges that included African agency. In the process, "the African diaspora has at times played a critical role in the making of its own alleged African 'base line' as well" (Matory 1999: 74). Fundamental to Matory's argument is the idea that Africa is "historically coeval" (ibid.) with the American cultures of which it is often said to be a traditional baseline. One might add that it is also coeval with the European cultures, which are often said to be more modern than it.

A different example might be the music industry in Latin America in the early twentieth century. Again, some standard assumptions about tradition and modernity might underwrite a narrative about modern recording techniques and their associated industry emerging in Europe and the United States, whence they spread into other areas, including the Caribbean and Latin America, creating local versions and hybrids. This is a powerful narrative—incontestable in some respects, particularly I suspect in relation to technical innovation. But it also backgrounds and misrecognizes the way this industry had a global dimension from very early on and actively constituted itself as an industry in the excursions its agents made into Latin America and other areas of the globe to record artists; in the interactions between the industry HQs in the United States and the local agents who doubled as retailers, PR men, and talent scouts; and in the creation of recording and production operations in Latin America.

The international recording industry began effectively with the founding of the Victor Talking Machine Company in 1901 and the Columbia Gramophone Company in 1903. From the early 1900s, these companies made "overseas recordings" by sending teams of representatives to foreign countries with stacks of wax blanks and recording machines which, although rudimentary, were quite portable. Using hotel rooms, local artists were recorded and the recorded blanks then used to make pressings back in the United States. Victor recorded thus in Mexico from 1905 and also Cuba. In the 1920s and 1930s, Victor established factories and studios in Santiago and Buenos Aires, not to mention Yokohama and elsewhere (Fagan and Moran 1986: 521). These companies also had networks of sales representatives all over Latin America and the Caribbean who, apart from selling phonograms and recordings, also recruited local talent. Such talent might travel at first to New York or Camden, NJ, and later to Buenos Aires or Santiago to make recordings.

The first recordings of Colombian musicians, made in New York around this period, give some idea of the multilateral movements of people and music (Wade 2000). In 1910 and 1917, Emilio Murillo traveled from Bogotá to make some recordings with the Columbia Gramophone Company and the Victor Talking Machine Company that included some *pasillos*, a two-step, a one-step, gavotas, polkas, and waltzes; he also directed a Colombian ensemble playing the national anthem. The Colombian duo Wills and Escobar were first recorded in about 1914 in Bogotá by the Victor Company. In the following years they toured Colombia, the Caribbean, and finally the United States, where in 1919 they recorded some Colombian songs, before visiting Mexico. Murillo and Wills and Escobar were also chosen to represent Colombia at the 1929 Exhibition of Seville. For all these artists, "Colombian" music was only part of their repertoire. The 1920 recordings of Colombian singer Jorge Añez included a Cuban bolero and a Mexican *ranchera*; his partner Alcides Briceño (actually a Panamanian) recorded foxtrots, marches, tangos, and Cuban *habaneras*.

This indicates the opportunist mix between the "national" and the "international" in the recording and marketing strategies of the record companies. On the one hand, national images were important. Songs were labeled with names that had definite national associations. Tango "came from" Argentina, rumba and guaracha from Cuba, ranchera from Mexico and bolero from Cuba—although Mexican singers and composers soon established themselves as leading exponents—while *bambucos, porros*, and *cumbias* came from Colombia. It is no accident that Murillo recorded the Colombian national anthem. On the other hand, the market and the recording techniques were highly transnational: musicians of many Spanish-speaking nationalities—not just Latin American, but Spanish and Canary Island too—played together, often in house orchestras that simply played sheet music sent in from different Latin American countries. Artists played many different styles alongside their "national" ones and the various styles were marketed all over Latin America. Some musical terms were not national. For example, the label "*canción*" (song) was often used. One singer of many *canciones* for Victor in this period was Juan Pulido, who was born in the Canary Isles, lived in New York, Cuba, and several other Latin American countries: this is typical of the transnational nature of this very generic style.

This indicates that notionally "national" repertoires were being produced in highly transnational circuits of production and exchange, both in terms of musical ideas and actual recordings. In the process, as Roberts (1979) has argued, "modern" music was being formed in these multilateral exchanges that not only defied national boundaries and categorizations but also conceptual divisions between tradition and modernity—even if, as always, people constantly deployed categorical distinctions of nation and modernity/tradition.

Conclusion

The question that inspired this collection asks when Latin America became modern. My answer is that it has always been modern, or as modern (and traditional) as anywhere else. This is not to deny differences between regions of the world and much less to deny power differences and economic inequalities. It is rather to question implicit teleological and scaling effects that place Latin America "after" modernity and as a "local" example of a global process that has its center elsewhere. It is to recognize the role of Latin America as *constitutive* of modernity, and thus coeval and colocated with it. It is one thing to recognize spatialized differences in technologies, in cultural practices, and social structures, and to enquire into the emergence and distribution of these—in a word, to investigate social change. It is another to channel that investigation into preformed spatiotemporal categories such as modernity and tradition or some iteration of center and periphery, which introduce subtle effects of teleology and scaling, even when the aim is to avoid such effects.

In my view, the notion of multiple modernities does not, in itself, avoid these problems. Anthropologists typically talk in terms of multiple modernities, which are formed by local people adapting globalizing cultural forms and producing hybrids in often unpredictable ways that maintain simultaneous dynamics of homogenization and differentiation. Wilk, among others, has argued that such hybrids often differentiate in similar kinds of ways: the globalizing forces set up basic structures of common difference, within which localities ring the changes (Wilk 1995). This puts the global and local back into a hierarchical relationship, in which the former holds the whip hand, but this particular version of the idea is not entailed by the basic notion of multiple modernities. My worry is about the way hidden effects of scaling and teleology remain in these ideas of modernities. This is not to say that the notion of modernity/ies should be abandoned. People not only use the concept all the time in everyday life, but put into practice projects based on the idea, projects that, like much human activity, attempt to create teleology, not only because they are themselves goal directed, but because they seek to establish such goals as models for social action. But this does not necessarily mean we should use the concept as an analytic one. Any analytic concept channels the attention in some directions and not others, but precisely because "modernity" comes charged with so much baggage, I think we need to be very careful about how we deploy it. All too easily, the concept reinforces dualisms and oppositions—between modernity and tradition, global and local—that need to be dissected and questioned. The critique of multiple modernities by Englund and Leach (2000) demonstrates how this happens very well.

Exploring concepts of modernity in the Chilean mining city of Antofagasta, Corsín Jiménez (2005: 171) concludes that "there is no place where one can locate 'modernity,' and in this sense it might be wiser to do without the term." His argument is that people use the term with reference to many different scales of change: changes in political economy, urban design, the politics of identity, the place of history in local identity, centralism and localism in national politics, a program of family values and practices of consumption. "We may call all these effective changes 'expressions of modernity' but this is all the heuristic value the concept will afford us. [. . .] The concept of modernity may speak to us about people's attitude to and aspirations for change. But the concept serves little more than a heuristic purpose, for it points to complexity, but does not elucidate it" (Corsín Jiménez 2005: 172, 173). In a sense, Corsín Jiménez is saying that the concept means too many different things for too many people to serve an analytic purpose. I agree, but would also add that hidden meanings also get smuggled into the analysis in ways that foreground some aspects of the situation being analyzed, while others remain in the background or are even made invisible. The danger lies precisely in the heuristic usefulness of the term: with such a handy hold-all concept, it is hard to disaggregate the various conceptual tools that are being brought to bear.

Note

1. Dunkerley (2000: 53) notes that Philip Curtin and Fernand Braudel both previously espoused notions of an Atlantic system of economic and intellectual exchange.

References

Berman, Marshall (1983). *All That Is Solid Melts into Air*. London: Verso.

Beverley, John (1999). *Subalternity and Representation: Arguments in Cultural Theory*. Durham: Duke University Press.

Castro-Gómez, Santiago (1998). "Latinoamericanismo, modernidad, globalización: prolegómenos a una crítica poscolonial de la razón," in Santiago Castro-Gómez and Eduardo Mendieta (eds.), *Teorías sin disciplina: latinoamericanismo, poscolonialidad y globalización en debate*. México: Miguel Ángel Porrúa.

Comaroff, Jean and John L. Comaroff (eds.) (1993). *Modernity and Its Malcontents: Ritual and Power in Postcolonial Africa*. Chicago: University of Chicago Press.

Corsín Jiménez, Alberto (2005). "Changing Scales and the Scales of Change: Ethnography and Political Economy in Antofagasta, Chile." *Critique of Anthropology* 25(2): 157–76.

Dunkerley, James (2000). *Americana: The Americas in the World, Around 1850*. London: Verso.

Dussel, Enrique (1995). "Eurocentrism and Modernity," in John Beverley, José Oviedo, and Michael Aronna (eds.), *The Postmodernism Debate in Latin America*. Durham: Duke University Press.

Englund, Harri and Leach, James (2000). "Ethnography and the Meta-Narratives of Modernity." *Current Anthropology* 41(2): 225–48.

Escobar, Arturo (2004). "Beyond the Third World: Imperial Globality, Global Coloniality and Anti-Globalisation Social Movements." *Third World Quarterly* 25 (1): 207–30.

Fabian, Johannes (1983). *Time and the Other: How Anthropology Makes Its Object*. New York: Columbia University Press.

Fagan, Ted and Moran, William (1986). *The Encyclopedic Discography of Victor Recordings, 1903–1908*. Westport: Greenwood Press.

Fontaine, Pierre-Michel (1981). "Transnational Relations and Racial Mobilization: Emerging Black Movements in Brazil," in John F. Stack (ed.), *Ethnic Identities in a Transnational World*. Westport, CN: Greenwood Press.

García Canclini, Néstor (1995). *Hybrid Cultures: Strategies for Entering and Leaving Modernity*. Trans. Christopher L. Chiappari and Silvia L. López. Minneapolis: University of Minnesota Press.

Gilroy, Paul (1993). *The Black Atlantic: Modernity and Double Consciousness*. London: Verso.

Graham, Richard (ed.) (1990). *The Idea of Race in Latin America, 1870–1940*. Austin: Texas.

Hardt, Michael and Antonio Negri (2000). *Empire*. Cambridge, MA: Harvard University Press.

Inda, Jonathan Xavier and Renato Rosaldo (eds.) (2001). *The Anthropology of Globalization: A Reader*. Oxford: Blackwell Publishers.

King, Anthony D. (1995). "The Times and Spaces of Modernity (or Who Needs Postmodernism?)," in Mike Featherstone, Scott Lash, and Roland Robertson (eds.), *Global Modernities*. London: Sage.

Kuper, Adam (1988). *The Invention of Primitive Society: The Transformation of an Illusion*. London: Routledge.

———(1993). "Post-Modernism, Cambridge and the great Kalahari Debate." *Social Anthropology* 1(1): 57–72.

Latour, Bruno (1993). *We Have Never Been Modern*. Trans. Catherine Porter. London: Harvester Wheatsheaf.

Matory, J. Lorand (1999). "The English Professors of Brazil: On the Diasporic Roots of the Yorùbá Nation." *Comparative Studies in Society and History* 41(1): 72–103.

Mignolo, Walter (2000). *Local Histories/Global Designs: Coloniality, Subaltern Knowledges, and Border Thinking*. Princeton: Princeton University Press.

Miller, Daniel, ed. (1995). *Worlds Apart: Modernity through the Prism of the Local*. London: Routledge.

Moreiras, Alberto (2001). *The Exhaustion of Difference: The Politics of Latin American Cultural Studies*. Durham: Duke University Press.

Pagden, Anthony (1982). *The Fall of Natural Man: The American Indian and the Origins of Comparative Ethnology*. Cambridge: Cambridge University Press.

Poole, Deborah (1997). *Vision, Race and Modernity: A Visual Economy of the Andean Image World*. Princeton, NJ: Princeton University Press.

Robbins, Joel (2004). "The Globalization of Pentecostal and Charismatic Christianity." *Annual Review of Anthropology* 33: 117–43.

Roberts, John Storm (1979). *The Latin Tinge: The Impact of Latin American Music on The United States*. New York: Oxford University Press.

Rosaldo, Renato (1982). "Utter Savages of Scientific Value," in Eleanor Leacock and Richard Lee (eds.), *Politics and History in Band Societies*. Cambridge: Cambridge University Press.

Sahlins, Marshall (1999). "Two or Three Things That I Know About Culture." *Journal of the Royal Anthropological Institute* 5(3):399–421.

Schwarz, Roberto (1992). *Misplaced Ideas: Essays on Brazilian Culture*. London: Verso.

Siegel, Micol (2001). "The Point of Comparison: Transnational Racial Construction, Brazil and the United States, 1918–1933." Ph.D. dissertation, New York University, New York.

Smart, Barry (1990). "Modernity, Postmodernity and the Present," in Bryan S. Turner (ed.), *Theories of Modernity And Postmodernity*. London: Sage.

Solway, Janet and Richard Lee (1990). "Foragers, Genuine or Spurious? Situating the Kalahari San Debate." *Current Anthropology* 31(2):109–46.

Stepan, Nancy Leys (1991). *"The Hour of Eugenics": Race, Gender and Nation in Latin America*. Ithaca, NY: Cornell University Press.

Stiles, Daniel (1992). "The Hunter-Gatherer 'revisionist' Debate." *Anthropology Today* 8(2): 13–7.

Taussig, Michael (1980). *The Devil and Commodity Fetishism in South America*. Chapel Hill: University of North Carolina Press.

—— (1987). *Shamanism, Colonialism and the Wild Man: A Study in Terror And Healing*. Chicago: Chicago University Press.

Therborn, Göran (1995). "Routes to/through Modernity," in Mike Featherstone, Scott Lash and Roland Robertson (eds.), *Global Modernities*. London: Sage.

Thompson, Grahame F. (2004). "Is All the World a Complex Network?" *Economy and Society* 33(3): 411–24.

Turner, Bryan S. (1990). "Periodization and Politics in Postmodernity," in Bryan S. Turner (ed.), *Theories of Modernity And Postmodernity*. London: Sage.

Wade, Peter (2000). *Music, Race and Nation: música tropical in Colombia*. Chicago: University of Chicago Press.

Wilk, Richard (1995). "Learning to Be Different in Belize: Global Systems of Common Difference." In Daniel Miller (ed.), *Worlds Apart: Modernity through the Prism of The Local*. London: Routledge.

Williams, Raymond (1988). *Keywords: A Vocabulary of Culture and Society*. London: Fontana.

Wilmsen, Ed and Denbow, James (1990). "Paradigmatic History of the San-Speaking Peoples and Current Attempts at Revision." *Current Anthropology* 31(5): 489–524.

Winant, Howard (1992). "Rethinking Race in Brazil." *Journal of Latin American Studies* 24: 173–92.

Wolf, Eric (1982). *Europe and the People without History*. Berkeley: University of California Press.

Chapter 3

Mid-Nineteenth-Century Modernities in the Hispanic World

Guy Thomson

Although by the turn of the twentieth century "modernism" (*modernismo*) had already been adopted by Hispanic writers and philosophers, with "modernization" entering common currency among U.S. social scientists during the 1920s, historians of Latin America have resisted using the terms "modernity" or "modernization" until quite recently.[1] Concerned with defining the peculiar and particular experience of early European-native encounters and precocious postcolonial histories, "modernization" and "modernity" have seemed too imprecise and too deferential to classic European or Anglo-American models of development models to be considered useful for understanding the complexity of Latin American history. Instead, historians of Latin America have favored a more selective and empirical vocabulary for describing processes and dichotomies seen as peculiar to the Iberian world: "hispanization," "luso-tropicalism," "mestizaje," "caudillismo," "civilization and barbarism," "development and dependency," "caciquismo," "apertura populista," "authoritarianism," and so forth. Yet it is hard to ignore the wealth of evidence that being and being seen to be "modern," particularly in emulation of the United States, was what elites in Spanish America had sought since long before political independence. In a challenge, then, to prevailing assumptions of Latin American exceptionalism, this chapter aims to illustrate how a methodology of comparative history can give the term modernity meaningful content that does not necessarily imply either the normativity or the teleology criticized by some other contributors to this book (particularly Wade). It does so mainly through exploration of two regional case studies: one from Mexico and one from the other side of the Atlantic, Spain. It argues that it is valid to claim that these areas of the Hispanic world became "modern" during the middle decades of the nineteenth century (1850s–70s), which was a period of accelerated exposure to external modernizing influences (see Dunkerley 2000). First, however, let us establish a broader context for these regional histories of precocious modernity by looking more closely at some of the diverse texts that have helped to shape the approach adopted here.

From Pioneering Anthropology to the New Global History

Scholars from other disciplines have not always shared the historians' caution about taking Latin American countries as laboratories for the incubation of a theory of modernization. Mexico was the first to receive such treatment: U.S. anthropologists working on rural communities there during the 1920s and early 1930s constructed models of culture change that still influence the way we study Latin American history (Clews Parsons 1930; Redfield 1930, 1934, 1941, 1950; Hewitt de Alcantara 1984; Tenorio Trillo 1999; Godoy 1977; Levy Zumwalt 1992; Deacon 1997). At first Robert Redfield and Elsie Clews Parsons approached their fieldwork in Tepoztlán and Mitla with a Boasian determination to eschew theory, particularly nineteenth-century biological theory, and to survey instead the cultural idiosyncrasies and complexity of small communities. Yet their very presence in rural Mexico—ethnographically uncharted since the first Franciscan missionaries—was evidence of a distinctively modern anxiety: the need to record, understand, plot, and even guide the outcome of economic and social changes, which were seen as imminent and inevitable.

In Mitla, Parsons was interested in observing the survival of a folk culture produced during an early period of intense acculturation between Spaniards and Zapotecs during the sixteenth century. But her 600-page ethnography also richly documents contemporary modernizing influences, such as the establishment of a new brass band to rival the one formed in the nineteenth century, the greater frequency of visits from government officials and candidates of political parties, the town's first baseball game, the first mechanized corn mill, motorized road transport, radio, film, newspapers, modern medicine, even the gum chewing and adoption of American clothing of her young Zapotec informer (and lover). Parsons allows the reader to deduce from these ethnographic snippets where Mitleños were heading. Like Parsons, Robert Redfield labored in Tepoztlán in 1927 to separate the

Indian from the Spanish cultural baggage, but eventually found the boundary between the "tontos," carriers of tradition, and the "correctos," vectors of progress, a much more profitable line of analysis. As might be expected from the son-in-law of Robert Park, Redfield soon abandoned Boasian antitheory and empiricism to construct models of cultural change to help calibrate the growing number of community studies, devising the idea of the "folk society," to refer to communities between the primitive and the modern, and the "folk-urban continuum," which was a spatial model for locating communities and individuals journeying along the road from the primitive, through the folk, to the modern.[2]

Although such models would be vulnerable to many of the criticisms outlined by Wade in Chapter 2, these ethnographic studies, with their appendices documenting informants' personal testimonies, have proved particularly valuable as benchmarks for later anthropologists, as well as evidence for historians of postrevolutionary Mexico searching for the presence of the nation-state in towns and villages.[3] Their approach has also provided inspiration for this chapter in emphasizing two factors that I see as crucial to thinking about modernity: first, the importance of subjective experience and, second, the insights to be gained from a local-level perspective.

Historians interested in what it means to be modern increasingly argue that ways need to be found to incorporate subjective factors into their analysis, in addition to the more "measurable" factors upon which they have tended to focus. One leading recent example is C. A. Bayly, whose definition of the "modern" in *The Birth of the Modern World 1780–1914* (2003) has been useful in conceiving this essay:

> ... an essential part of being modern is thinking you are modern. Modernity is an aspiration to be "up with the times." It was a process of emulation and borrowing ... between about 1780 and 1914, increasing numbers of people decided that they were modern, or that they were living in a modern world, whether they liked it or not ... the nineteenth century was the age of modernity precisely because a considerable number of thinkers, statesmen, and scientists who dominated the ordering of society believed it to be so. It was also a modern age because poorer and subordinated people around the world thought that they could improve their status and life-chances by adopting badges of this mythical modernity, whether they were fob watches, umbrellas or new religious texts. (Bayly 2003: 10–11)

Bayly's emphasis upon subjective experience, particularly the imagination of people of the "middling sort," matches the approach

adopted later. His book does, however, raise some further methodological issues that are also pertinent here, albeit in less positive ways.

The Absence of "las Españas"[4] in Bayly's
The Birth of the Modern World

Although this remarkable book aspires to be a global history of the
long nineteenth century (1780–1914), it is mostly an exploration of
patterns of interconnectedness and interdependence—political, economic, cultural, and ideological—between Europe and the Orient
(Middle East, India, China, and South East Asia, with glances at
Africa and the Antipodes). The Americas (Bayly looks chiefly at the
United States and the Caribbean, very little at Latin America) are
brought in to explain the foundations of European global economic
dominance and colonialism during the period that he calls "archaic
globalisation" (between 1600 and 1800). But the book is mainly
about the endurance and intensification of European dominance in
Africa and Asia during the nineteenth century. The Hispanic world,
for the most part already postimperial and postcolonial by the third
decade of the nineteenth century, hardly features among the book's
600 pages.[5]

Global histories are necessarily selective. Bayly's omission of the
Hispanic world can perhaps be explained by the difficulty of fitting
histories of former European colonies into a framework designed for
exploring how the populations of new colonies or dependencies in
Asia and Africa absorbed and contested influences from Europe
(a process that had already happened in Spanish and Portuguese America
during the sixteenth and seventeenth centuries). From 1810, Spanish
America and Brazil freed themselves from Europe but kept the social
and ethnic hierarchies of colonialism. On the few occasions Spain or
Spanish America are referred to, the image is one of the immutability
of structures set during the period of archaic globalization—the predominance of latifundia, dependence on primary exports, extremes of
wealth and poverty, power and powerlessness, Catholic religious
uniformity—structures that allowed for little contestation from subalterns and afforded little room for modernity. Only in the Latin
America of the 1980s can Bayly discern comparable cultural and political assertiveness among Latin America's indigenous population to the
kind of resistance to European colonialism that he charts in Asia and
Africa a century earlier. Of course, he would have found plenty of
evidence of comparable subaltern resistance to postcolonial power
structures in Latin America during the second half of the nineteenth

century. But this would have greatly complicated his analysis. Social, ethnic, and cultural boundaries between Europeans and non-European in Africa and Asia, even after a century of European dominance, are much easier to discern than in Latin America, where ethnic and social hierarchies are more complex and hybrid, and resistance is harder to observe and politically less conclusive.[6]

Hence, while Europe, Africa, and Asia lived through the intimate global convergences and intensification of connections of the "Age of Capital" and the "Age of Empire," the Hispanic world experienced the opposite: the disintegration of its former territorial, political, religious, and economic unities (Hobsbawm 1988–98). Of course, "Hispanic capitalism" (a concept coined by Loveman, 2001) survived the political collapse of "Las Españas," and continued to provide plenty of space for localized modernization. But "Hispanic capitalism" was distinguished by its parochial character, its penchant for protectionism, its accommodation to quite low levels of consumption of predominantly agrarian and unequal societies, and a resistance to transforming social relations and structures, cultural practices, and religious beliefs.

Perhaps we are fortunate that Bayly chose not to follow the example of David Landes who includes lengthy sections on the Hispanic world in his 1998 global economic history, *The Wealth and Poverty of Nations*. A book that gives the new global history a bad name, bereft of any engagement with recent historiography, Landes mocks Hispanic intellectual obscurantism and economic backwardness and seems intent only to confirm a Black Legend-inspired view of Latin America. A paragraph will suffice:

> . . . the history of Latin America in the nineteenth century was a penny-dreadful of conspiracies, cabals, coups and countercoups—with all that entailed in insecurity, bad government, corruption, and economic retardation.
>
> Can any society long live in such an atmosphere? Or get anything done on a serious, continuing basis? The answer is that these were not "modern" political units. They had no direction, no identity, no symbolism of nationality; so no measure of performance, no pressure of expectations. Civil society was absent. At the top, a small group of rascals, well taught by their earlier colonial masters, looted freely. Below, the masses squatted and scraped. The new "states" of Latin America were little different, then, from Asia's autocratic despotisms, though sometimes decked with republican trappings. (Landes 1999: 313)

To his credit Bayly dispels comparable European stereotypes that presented the Orient as unmitigated despotism. But armed only with

Peter Bakewell's general history of Latin America—part of the same "Blackwell History of the World" as *The Birth of the Modern World*— he is unable to do the same for the Hispanic world.

The Birth of the Modern World is also intended in part as an antidote to what Bayly sees as a rejection by postmodern and postcolonial historians of "grand narratives" and their tendency to wallow in the local so as to "to recover the 'decentered' narratives of the people without power" (Bayly 2003). He does admit, however, that "histories of the experience of individuals and groups isolated from the main centres of production of history" can be important:

> The marginal has always worked to construct the grand narrative as much as the converse has been true. Especially before the mid-nineteenth century, it was common for people on the "fringes" to become historically central. Nomads and tribal warriors became imperial generals. Barber surgeons became scientists. Dancing women became queens. People easily crossed often flexible boundaries of status and nationality. Historical outcomes remained open. . . . Yet it is difficult to deny . . . the importance of the weight of the change towards uniformity over the "long" nineteenth century.
>
> . . . post-modernist works usually conceal their own underlying "meta-narrative," which is political and moralising in its origins and implications . . . many of these accounts appear to assume that a better world might have evolved if such historical engines of dominance as the unitary state, patriarchy or Western enlightened rationalism had not been so powerful. All histories . . . even histories of the fragment are implicitly universal histories. (Bayly 2003: 9)

Although the following accounts unashamedly qualify as "histories of fragments," even as "decentered narratives of people without power," they have been selected not from any conscious moralizing or political agenda, but because they provide a sound comparative basis for testing the meaning of the term modernity in two locations that are conventionally assumed, as exemplified earlier, to be unusually resistant to modern ideas and practices.

The region of Latin America to be examined is the Puebla Sierra of East-Central Mexico during the period between the Liberal revolution of Ayutla of 1854—regarded by Clara Lida as Mexico's 1848 (Lida 2002)—and the early 1890s when the conservative-liberal regime of Porfirio Díaz (1876–1911) suppressed a rash of regional and local uprisings led by the dictator's former Liberal-patriotic companions at arms who in the middle years of the century had helped construct a more yielding state. The region to be examined in Spain is the mountainous borderland of the provinces of Córdoba, Málaga, and Granada, between the progressive Liberal revolution of 1854–56— the "Bienio Progresista" regarded as Spain's 1848—and the early 1880s, six years into the conservative-liberal Bourbon Restoration, when the democratic optimism of the 1850s and 1860s degenerated into nihilism, mass arrests, and deportations. Thus the rationale for the comparison is that these two regions were both during the same period (1850s–80s): at an early stage of exposure to technology and capitalism; relatively far from the centers of political power in their respective countries; undergoing a political shift from liberalism to authoritarianism; and experiencing the rise of middle and working classes. The evidence from these two regional case studies reveals comparable struggles of people from the middle classes who shared a common set of ideas about democracy and progress, admired the same contemporary democratic celebrities, and succeeded in devising ways of gathering popular followings at a time when their opponents were finding it hard to attract support. These mid-nineteenth-century manifestations of modernity in the Hispanic world are explored thematically later under the following headings: (1) democratic optimism; (2) cultures of consumption; (3) new forms of sociability; and (4) the politicization of solidarities.[7]

Imminent Happiness: Mid-Nineteenth-Century Democratic Optimism

En la marcha del entendimiento humano, un año ni dos forman mas que unos puntos casi imperceptibles á la vista, y hasta las grandes épocas históricas vienen á ser únicamente lunares diminutos que apenas se divisan en la inmensidad del gran espacio que abraza lo pasado, lo presente y lo que está en el porvenir.

Pero la época que atravesamos, por su carácter analítico, de examen y de crítica, por su espíritu eminentemente reformador, por la gravedad de las cuestiones que se tratan de resolver de modos diversos, reasume un siglo de combates entre el bien y el mal, entre el espíritu y la idea, entre los que fue y lo que habrá de ser.

—Pérez Luzaró 1853: 8

From the 1840s people throughout Europe and the Americas believed they had entered a new age of technological change and human progress unlike anything that had gone before. In particular, vocal sections of an emerging middle class believed that they possessed the "ideas of the century" and were acquiring the social and political influence to become handmaidens of this new age of liberty, prosperity, and international brotherhood. In the end, the democratic optimism of the middle class and its belief in being the anointed agents of social emancipation proved illusory. But for 30 or so years, Democrats in Spain and radical Liberals in Mexico believed that they were the natural custodians of ideas that would transform states and societies, just as their opponents—Moderate Liberals and Neo-Catholics in Spain or Conservatives in Mexico—feared their own ideas might be heading for an early exit, with the utopian prophecies of their enemies realized sooner rather than later.[8]

A variety of circumstances encouraged this democratic optimism. Sustained population growth, accelerated industrial growth, and agricultural transformation, the revolutionary shortening of distance brought by the railway, the steam ship and the telegraph, combined to create a culture of consumption and material optimism. The expansion of the postal service, the greater ease of distributing national newspapers, the greater speed in gathering news of local and international events, contributed to a quantum leap in up-to-date knowledge of national and world affairs among expanded reading publics. The content of news also seemed to confirm both the onward march of liberty and progress and the fragility of the ancien régime: the faltering of European dynasties (the Habsburgs in Austria and Italy and the Bourbons in Sicily and Spain), the triumph of the Union in the American Civil War and the abolition of slavery, the triumph of Mexico's Liberals over Maximilian's Second Empire, Italian Unification and the challenge this posed not only to the temporal power of Pius IX but to countries such as Spain where Catholicism was regarded by many as synonymous with the nation, subaltern challenges posed to European power in new colonial territories (the "Mutiny" in India and the Taiping rebellion in China), and the failure of several new colonial ventures (Russia in the Caucasus and France in Mexico).[9]

Adding to this sense of imminent dynastic and ancien régime meltdown, short-term economic downturns and mid-century subsistence crises (particularly gravely felt in Spain), often accompanied by riots and labor militancy, highlighted the fragility of traditional political structures and the corruption, incompetence, and decadence of ruling elites. Finally, although a modern, more centralized, and ultimately much stronger state was taking shape during the same generation, the creators of this new state—Mexico's Conservatives and Spain's Moderados—failed in their attempts to endow it with the legitimacy or political strength it needed to command obedience.[10] The Catholic Church in Spain and Mexico was in no condition to provide this prop, with its finances and personnel ravaged by the *desamortización*. In both countries, the Church was slow to adapt itself to the new secular political environment, social formations, and modern communications, or to find a way to compete with the attractiveness of democratic and secular prophecies.

The vacuum left by the retreat of the Church drew in a host of secular social and democratic doctrines, particularly after 1848 with the formation in Spain of the Democrat Party and the emergence in Mexico of radical liberalism in the wake of the American War (1847–48). Although the nationalist agenda of 1848 had little relevance in the Hispanic world, the internationalist, Iberianist (unification of Spain and Portugal), pan-latinist, and democratic ideas that flourished during the 1850s and 1860s offered Spain a way out of intellectual and political isolation. Spaniards returned from periods in foreign exile, or from spells in overseas *presidios*, with Owenite, Fourierist, Christian Socialist or Mazzinian convictions, and a determination to create a new order based upon free association of citizens, workers, and women, for education, savings, collective bargaining, preparation for entry in the electorate, and religious practice. Along with Lincoln, Kossuth, Garibaldi, and Mazzini, their principal hero was Benito Juárez who in 1867 freed Mexico from Europe's worst despot. European exiles brought similar ideas and experience to Mexico (Lida 2002; García Cantú 1974). In December 1865, Mazzini even persuaded Juárez to accept Roberto Armento, Chief of Garibaldi's artillery in 1860–61, to serve in the Liberal insurgent army.[11]

Hence, from the 1840s, the Hispanic world experienced a revival of democratic ideas, but directed at a wider section of the population than the constitutional liberalism of the early nineteenth century. Democratic ideas evoked patriotism but also stressed international fraternity, denounced class privilege and political exclusiveness but also urged class harmony, condemned religious intolerance and Catholic reaction but also tapped into Christian ideas of virtue, community, brotherhood, and self sacrifice; promoted economic freedom and individual property ownership but also preached sobriety, frugality, cooperative endeavor, and cheap credit. Above all and in contrast to earlier liberal constitutional ideas, democratic ideas were broad in their appeal. Democratic catechisms read like checklists of reasonable things to be done.[12] Only the selfishness of discredited ancien régime elites and political institutions stood in the way of their fulfillment. It was precisely the moderate and commonsense quality of the ideas and the middle-class respectability of the propagandists that made Conservatives so fearful of "Democracy."

During the 1850s and early 1860s, Conservatives in Mexico and Moderate Liberals in Spain stared into an abyss populated by a noisy, bearded rabble that was already behaving as if the democratic future was theirs. In March 1864, the Conservative newspaper *El Espíritu Público* published a report from its correspondent in Alhama (Granada):

> They write to us from Alhama telling us that we are in an indescribable epoch, in which it seems that men from all the hierarchies are determined to contribute so that our period of history can be written in letters of blood. The revolution, our correspondent adds, can be seen growing here, as in all other parts, in a terrible way, and it is certain that we are in the same position, exactly the same position, as in the days preceding the Loja uprising (1861). Meetings are held in every place; democratic and socialist newspapers are read in great choirs in the streets; those who profess these ideas use the beard as an emblem, each one believing himself to be a minister or a revolutionary. God help us survive this summer in peace! Spain, with any doubt, finds itself in a period very similar to the century of the destruction of ancient Rome, in which ambition, luxury and pleasure destroyed what would never figure again.[13]

Let us now look more closely at the regional and local level to explain how this democratic optimism took hold.

Cultures of Consumption

Both regions experienced rapid economic and demographic growth from the 1850s to the 1880s (and later). This growth was possible even without significant transport improvements, thanks to mules, wagons, and tumplines (the railway did not reach the Puebla Sierra until the eve of the Revolution of 1910 and an unbroken railway connection did not connect Granada with Málaga until the 1880s). Growth was fueled by a combination of the privatization of corporately held and common lands, the expansion of new cash crops and economic activities, and the growth of consumer demand. In Mexico, coffee and aguardiente became the great money-spinners, attracting entrepreneurs from the tropical lowlands and the cold plateau into the Sierra. In the Granada–Málaga highlands, economic growth resulted not only from the opening of new land for olive, wine, and wheat production, horse and sheep rearing on former Crown, seigniorial and municipal commons, but also from the mechanization of the woollen and paper industry (in Loja and Antequera), which attracted technicians—with democratic ideas—from France, Catalonia, and Valencia.

Economic development and the opportunities for opening new land for commercial agriculture increased the wealth of the existing elite, facilitated the expansion of a middle class, encouraged the emergence of a newly propertied peasantry (that emulated and hoped to join this middle class), and increased the wage-earning and bargaining power of the landless laboring population (*jornaleros*). All classes began to consume more. This was reflected in house building, tiles replacing thatches, the standardization among the middle class and middle peasantry of modern, generally dark, clothing (effective for disguising wealth differences), the use of stiffened felt hats and leather shoes. While for most people the revolution in consumption was a modest affair, the enjoyment of conspicuous luxury now became a possibility among those who benefited most from the land booty and the commercialization of agriculture.

Especially in small towns, conspicuously wealthy people had traditionally been expected to be conspicuously active in providing charity and ritual leadership. But in Liberal Spain, many of the wealthy families resided away from their hometowns, in provincial capitals or in Madrid, for much of the year. Even when they returned they often chose to reside away from rowdy and lawless agro-towns on their rural properties.[14] Other members of the newly propertied landed elite preferred to disguise their wealth, concerned to keep their tax assessments low (the blandishments of the high property requirements to vote were often not enough to compensate for the fiscal risk of being rich). This habit, along with elite absenteeism, combined to encourage a democratic uniformity and sobriety of taste in southern Spanish agro-towns. The retreat of the new landed elite into the private sphere, a paradoxical consequence of those ages of rapid material advance and class formation, also magnified the influence of members of a nascent professional middle class whose livelihood (often as municipal employees) required their conspicuous presence in towns: doctors, pharmacists, veterinarians, and lawyers, who formed the core of democratic sociability.

In Mexico, where the influx of wealth was altogether a novelty in the Sierra regions and where the new rich were resident rather than absentee, excessive wealth tended either to be disguised, or used, in the case of Liberals, for civic and transport improvements, equipping bands and schools, political patronage and warfare, or, later in the century, following the entente between Church and state, for church building and the promotion of religious festivals that tied the coffee bourgeoisie in with dependent indigenous communities.

New Forms of Sociability

The 1850s witnessed in both Spain and Mexico the introduction of compulsory primary schooling, the displacement of the Catholic Church from the center of ceremonial life, and the development of secular associational life around cafes, taverns, reading rooms, casinos, and Sunday National Militia (Spain) and National Guard (Mexico) musters. The continuing financial and institutional weakness of the Church, the legacy of ecclesiastical disentailment and Liberal anticlericalism, left a vacuum in which secular and democrat ideas and forms of association faced little resistance. In Mexico, the exclusion of the Church from the public sphere was enforced by the Reform Laws. In Spain, the clergy was particularly weakly represented in rural areas, especially in the South.[15]

The secularization of associational life and the public sphere was dramatized in both countries by the proliferation of musical societies and wind bands whose repertoire echoed the global taste for Italian opera, Austrian waltzes, and Scottish dances. Of course, these bands could be easily trained in sacred music and deployed in religious festivals, and frequently were. But they tended to be municipally funded, with band masters recruited from the army, bringing with them a martial and patriotic repertoire, such as the Hymn of Riego in Spain, or in Mexico, melodies and lyrics lambasting the French and the Conservatives. The aural conquest of the external space of towns by the brass band was matched in the interior space of cafés and taverns by rowdy songs celebrating, in Spain, the exploits of Garibaldi, Lincoln, and Juárez, and in Mexico, liberal and patriotic victories.

If secular cults of democratic heroes helped fill the vacuum left by the retreat of Catholicism, the spiritual thirst and social aspirations of sections of the population were addressed during this period by Protestant missionaries. In Mexico, Methodist missionaries were promoted by the Juárez government; in Spain, Protestants and members of the Bible Society entered Málaga and Granada clandestinely from the Balearics or from Gibraltar. Both regions experienced a significant growth of small but influential Protestant congregations. Membership of these fledgling congregations correlated closely with radical liberal and democratic political affiliation and sympathies.[16] In March 1869, a Presbyterian minister from Málaga visiting the inland textile city of Antequera had to be rescued from a mob demanding New Testaments in Spanish, the 300 copies he had brought from a Málaga printer having been snapped up immediately upon arrival at the railway station.[17]

Overall, the secularization of the public sphere, the omnipresence of news (blanket coverage of the Italian Risorgimento between 1858 and 1865 was interpreted by Spain's Democrats as a portent for country's own long awaited regeneration), tended to demoralize Conservatives and the clergy who retreated into the private sphere, and to encourage radical Liberal and Democrats who felt that the future was theirs for the taking.

The Politicization of Traditional Solidarities

Little would have come of all this democratic exaltation had it not been for the ability of "men of ideas," generally from the middle class or the aspiring middle class, to transform democratic ideas and modern sociability into political influence and mass followings. In Mexico, Liberal leaders from the Puebla Sierra contributed significantly to the triumph of Liberals over Conservatives between 1857 and 1861, Mexicans over Europeans between 1862 and 1867, and radicals over moderate Liberals between 1868 and 1876. Masonic *mestizo* village schoolteachers, from modest Sierra backgrounds, became Puebla state governors under the first Díaz administration.

In Spain, the Málaga–Granada borderlands produced no Republican leader of national stature (indeed Loja was the home of the Conservative Ramón María Narváez, savior of Papacy against the Rome Republic in 1849 and tormenter of Democrats between 1848 and his death in 1868). However, the Democrat Party, along with "Young Spain," were founded in the Madrid residence of noblemen from Antequera, along with Loja, the principal city in the region. Workers and day laborers in Loja and Antequera, which were industrial as well as agricultural centers, were targeted by Democrat leaders and mobilized in a series of (abortive) uprisings, the first in the summer of 1857, the last in autumn of 1869. Among the leaders of these uprisings could be found veterinarian-blacksmiths, newspaper sellers, and bookshop owners, small and even some large landowners, pharmacists, doctors, lawyers, Protestant preachers, the occasional returnee priest from America, master weavers, estate foremen, former officers and n.c.o.s of the National Militia, even Gypsy chieftains. In spite of their failure, these democratic uprisings left a legacy of association, memory, and myth that nourished later nineteenth-century republican and anarchist sociabilities, serving as an inspiration to the revival of republicanism during the 1920s and 1930s, and to the revival of the Left in the region following the death of Franco in 1975.

In Mexico, the political success of Sierra Liberals lay in their ability to organize the predominantly indigenous peasantry into locally controlled and financed companies of the National Guard. This was achieved by promoting literate and nonliterate Nahuas to positions of military command and to municipal presidencies (even to District government in the case of the principal Nahua military leader, Juan Francisco Lucas, whose preeminent influence as the Liberal "Patriarch of the Sierra" and leader of the indigenous population endured until his death in 1917). The enduring political success of Sierra Liberals encouraged the emergence of an ethnically hybrid modern sociability in towns such as Xochiapulco, where a Protestant congregation coexisted with a Catholic cult of the Saints, where girls and boys acquired non-Indian surnames and precocious literacy in Spanish, and were exposed to a cosmopolitan republican political culture that revered Cuautemoc, Hidalgo, Morelos, and Juárez, but also Washington, Lincoln, and Garibaldi (and yet in 1927 continued to communicate with each other and their families in Nahuatl).[18]

Hence, the National Guard enabled Sierra Liberal elites to assert themselves in state and national politics. It also provided a medium for Indian and peasant communities to negotiate with local elites and to protect their autonomy. In the mountains between Málaga and Granada, Democrats at first attempted to put the National Militia (Spain's equivalent to Mexico's National Guard) to similar ends. However, the heyday of the National Militia had already passed (in the struggle against the French during the 1810s and early 1820s and following the end of absolutism between 1837 and 1843). In the aftermath of the Bienio Progresista of 1854–56, when the National Militia had briefly been reconstituted, and during the Sexenio Revolucionario of 1868–74, when it reappeared as the Volunteers of Liberty, these municipally armed and controlled militias, commanded by Democrats and manned by artisans, workers, and peasants, were often violently demobilized by centralizing and repressive regimes. Hence, the Progressive Liberals, Democrats, and Republicans grew to regard the National Militia more as a liability than a means for attaining their political objectives.

Spain's Democrats, in any case, aspired to transform liberal politics from its patrician mold, exemplified by the patriotic National Militia, to a politics representing all Spaniards. The means they adopted for consolidating support among the wider population were clandestine techniques imported from Italy during the struggle against absolutism during the 1820s. Between 1855 and the late 1870s, Democrat leaders in the Málaga–Granada borderlands promoted membership of *carbonari* societies among agricultural day laborers and factory workers.

Using the same decurial structure adopted by Protestant evangelists at the same time, Democrats met with extraordinary success. Tens of thousands of industrial and agricultural laborers, and broad swathes of the lower middle and middle classes swore allegiance to these societies. However, they failed to translate mass membership of *carbonari* associations into successful uprisings or more than momentary electoral success.

Just as Liberals in the Puebla Sierra learned that they could tailor their military organization to the needs of Indian communities in exchange for political support, Spanish Democrats learned how to capitalize on the existing solidarities of Andalucian day laborers and factory workers. In particular they could take advantage of the preference of *jornaleros* to work in large gangs under leaders who were fellow workers. Democrats also benefited from the tradition of "unión": the habit of withdrawal of labor or going slow in order to exact better pay and work conditions, and the tradition of cold shouldering fellow workers who broke rank.[19]

Democrat success in recruiting workers into *carbonari* societies was also aided by the day laborers' habit of gathering in the main square or on the edge of towns at the start of the day. Traditionally, *jornaleros* were met here by employers who, even during slack periods of the year but especially during periods of drought, harvest failure, and unemployment (with a little prodding from the municipal authorities), would share out the workers, take them back to their estates and put them to work on repairing walls and building roads in exchange for subsistence. This customary practice, known as *alojamiento*, gradually died out over the 1850s and 1860s, as the gravity and frequency of subsistence crises increased and landowners refused to bear the social cost. Democrats stepped in with promises of credit cooperatives, the subdivision of illegally acquired municipal land, and of using democratically controlled local government as a source of employment and social security.

Democrat leaders also took advantage of a growing tavern culture, a reflection of improving living standards and increased alcohol consumption. Here they could find popular leaders of the criminal underclass whom they would recruit and deploy in such tactics as the sending of threatening letters ("anónimos"), kidnappings and incendiary activities, considered by some Democrats as a justified response to the repressive tactics of their political opponents, or as a way of punishing the new rich for their illicit appropriation of municipal land. Democrats also succeeded in swearing in to *carbonari* membership entire rural hamlets that were seeking release from seigniorial jurisdiction and promotion to municipal status (a common incentive for Indian communities in Mexico to join the National Guard was to achieve the independence of *pueblos sujetos* from oppressive *cabeceras*).

Finally, Democrats benefited from the requirement that each head of a decurial section of the *carbonari* society purchase a democratic newspaper to be read out aloud at break time. It was probably no coincidence that the peak of *carbonari* membership coincided with the Italian Risorgimento (1858–65), news of which Democrat organizers and laborers in the region found particularly relevant to their own situation. For their part, schoolteachers in the Puebla Sierra during the 1870s and 1880s confected a heroic, patriotic, and praetorian narrative from their own experience in the struggle against the French Intervention, embellished with memories of Cuautemoc and Xicotencatl, heroes of the struggle against the Conquistadores.[20]

Conclusion

Even in Mexico, where Conservatives and the Catholic Church found themselves out in the cold until the late nineteenth century, radical liberal optimism, and the belief that democratization and economic progress would continue to march hand in hand, began to be tempered after the restoration of the Republic in 1867. During the second presidency of Porfirio Díaz (1884–88), Conservative Liberals prevailed over Radicals (Hale 1989). The growing power of the state, and its commitment to economic progress at all costs, transformed attitudes toward subaltern assertiveness. The National Guard was demobilized in 1888. The pockets of radical liberal and nonconformist sociability formed during the 1860s and 1870s remained just pockets. The image of the Sierra regions of southeastern Mexico was transformed from that of the home of liberty and liberal-patriotic resistance against Conservatives and Europeans, to one of *caciquismo*, the despoliation of Indian community lands and general misery of the Indian population. By the 1890s, the new cultural symbols in the countryside had become, in the Puebla tableland, palatial hacienda buildings refurbished in garish eclectic European styles, and in the wealthy coffee towns of the Sierra, newly constructed Gothic churches.

In Spain, the "Sexenio Revolucionario" (1868–74) and the First Republic (1873–74) proved to be profoundly disillusioning for Democrats (now Republicans), whose mood of despair was further deepened by news of the Paris Commune. In truth, Democrat optimism was already dented well before the "Glorioso" of September 1868. The establishment in London in September 1864 of the International Working Mens' Association and the rapid growth of support for the International in Spain ended the monopoly that Spanish Democrats had enjoyed as the sole party able to point the way toward a new age of international fraternity and progress of mankind.

As worrying for Spanish Democrats during the late 1860s was the emergence of Conservative Liberalism, the ideology of the reconstituted landed elite, beneficiaries of the *desamortización*. During the constitutional chaos of Sexenio Revolucionario, young Conservative Liberals quickly learned the value of the "partido de la Porra" (the "party of the club"): low-level political violence that could be used to intimidate voters and make elections. The Paris Commune gave all parties an excuse to denigrate Republicans, to ban their meetings, and to forbid them from participating in elections. Faced with political exclusion, Republicans staged their last, fruitless armed rebellion in October 1869, attracting an impressive 40,000 young republicans into the field, but achieving only a store for future of memories of defiant failure. The Bourbon Restoration in 1874 ensured the continuation and perfection of a system of centrally directed electoral violence and exclusion (that adopted the American label of "caciquismo"). The Restoration also offered conditions propitious for a Catholic revival.

C. A. Bayly traces this pattern of transformed gentries, revived aristocracies, restored and reinvented monarchies throughout Europe and the wider world from the 1870s, and points to its fatal legacy in a violent twentieth century (Bayly 2003: 395–431). "The Reconstitution of Social Hierarchies" and the "The Transformation of Gentries" did not mean, of course, the end of modernity. Far from it, ever greater numbers of people throughout the world grew to think of themselves as modern, to aspire to "keep up with the times," to emulate and to borrow ideas and tastes, generally originating in Europe. But it did mark the end of a period in the mid-nineteenth century when groups of people throughout the world had believed that ideas of class harmony, international democratic solidarity and fraternity, a gradual leveling of social hierarchies and the expansion of the middle class, the abandonment of caste and class privilege, the ending of slavery, the replacement of monarchy with republics, were self-evident truths, virtues to be propagated, and marks of modernity.

Notes

1. Literature specialists approach the questions of modernity and modernization with greater certainty and chronological precision. Gerald Martin writes (1989: 365):

   The modern world had dawned in about 1870 and the Latin American poets began to unite and to prepare for it. Modernity itself, like Latin American modernization, arrived after 1917, at which point novelists began to engage with it, as poetry met prose and literature met criticism all over the West. Latin America's full entry into the modern world was patently visible after 1945–celebrated by Octavio Paz in 1950–albeit on radically unequal terms.

   For an explicit application of a modernization model to the social history of postrevolutionary Mexico, see Vaughan (1997).
2. Redfield (1940 and 1947); for a critique of Redfield's "folk society" idea and of his study of Tepoztlán, see Lewis (1951: 427–48).
3. Oscar Lewis worked in Tepoztlán during the 1940s, when Redfield returned to the Yucatan to check up on Chan Kom's journey to "progress" (Redfield 1950); Alan Knight (1994) and Mary Kay Vaughan (1997) both make extensive use of community studies to illustrate the depth of political and cultural change on the local level following the Revolution. See also Fallaw (2002: 645–84).
4. Not until after the Bourbon restoration with the new constitution of 1876 was Spain referred to officially in the singular (Esteban 1981: 177).
5. Spain is noted only for its credulous and fanatical Carlist peasantry. Portugal is not mentioned at all. Mexico, which Bayly locates in Central America rather than in North America, receives intermittent attention. The burial and reburial of Antonio López de Santa Anna's leg is mentioned on two occasions in order to illustrate the desperate lengths to which caudillos would go in order to seek legitimacy. Bayly also notes continuities between Mexico's patriotic peasant-backed and Garibaldi-led (*sic*) struggle against Louis Napoleon in the 1860s and the nationalist and "messianic" (*sic*) revolution of 1910, but misses a chance to draw parallels between these events and early nationalist and anticolonialist movements in India and Southeast Asia. Bayly (2003: 3–4, 141, 147, 161, 376).
6. "It was the capacity of European companies, administrators and intellectual actors to co-opt and bend to their will global networks of commerce, faith, and power that explain their century-long dominance. This is not so much a theory of collaboration, as one of subordination . . . it was some years beyond the terminal date of this book (1914) that this European dominance began to flake and decay over much of the colonial world . . . the 1930s in India and China, the 1950s and 1960s in Africa and the 1980s in the Soviet Empire and the Latin American world, as native and indigenist movements began to emerge" (Bayly 2003: 476).

7. The Puebla Sierra case study is based upon Thomson with LaFrance (1999) and Thomson (1994). The Spanish case study is based on *Thomson* (2001) and a book project reaching completion, *Before Anarchism. The Birth of Modern Politics in Southern Spain, 1849–1879.*

8. For democratic ideas in Spain and Mexico, Demetrio Castro Alfín, "Unidos en la adversidad, unidos en la discordia: el partido demócrata, 1849–1868," in Nigel Townson, ed., *El republicanismo en España (1830–1977)* Madrid, 1994; Antonio Eiras Roel, *El Partido Demócrata Español (1849–1868)* Madrid, 1961; José María Jover, "Conciencia burguesa y conciencia obrera en la España contemporánea" in José María Jover, *Política, Diplomacia y Humanismo Popular en la España del siglo XIX* Madrid, 1976, 45–82; Jordi Maluquer de Motes, *El socialismo en España 1933–1868* Barcelona, 1977; Clara E. Lida, *Antecedentes del movimiento obrero español (1835–1888)* Madrid, 1973; Clara E. Lida, *Anarquismo y Revolución en la España del XIX* Madrid, 1972; Gastón García Cantú, *El socialismo en México (siglo XIX)* Mexico, 1974; John Mason Hart, *El anarqismo y la clase obrera mexicana (1860–1931)* Mexico, 1980; Alan Knight, "El liberalismo mexicano desde la Reforma hasta la Revolución (una interpretación)," *Historia Mexicana* XXXV, 1985, 59–91; Guy P. C. Thomson, "Popular Aspects of Liberalism in Mexico, 1848–1888," *Bulletin of Latin American Research* X, 1991, 265–92; Clara E Lida and Carlos Illades, "El anarquismo europeo y sus primeras influencias en México," Unpublished paper, El Colegio de México, 1999.

9. Citing the dramatic events in Italy, the crisis in the German federation, the antecedents of the abolition of serfdom in Russia, George Weill observes how between 1859 and 1861 "a change in continental Europe brought the press into politics before it had been given the freedom to do so . . . All of these events were too grave to be able to hide from the public. Newspapers began to expose them and comment on them . . .," George Weill, *El periódico. Orígenes, evolución y función de la prensa periódica* Mexico, 1962, 162; in Spain, Francisco Giner de los Ríos, Krausist propagandist of "enseñanza libre," later identified ". . . the ten years between 1860 and 1870—if one has to fit arbitrary limits—as one of waking up from the old drowsiness (*modorra*) into the hum of modern European thought and to the problems and new postulates of its philosophy . . .," both cited in Jover (1976: 340–41).

10. For early state building in Mexico under the Conservatives, see Will Fowler, *Tornel and Santa Anna: The Writer and the Caudillo, Mexico 1795–1853.* Westport, Conn., 2000, and in Spain, F Cánovas Sánchez, *El moderantismo y la Constitución española de 1845,* Madrid, 1985, and Nelson Durán, *La Unión Liberal y modernización de la España Isabelina. Una convivencia frustrada, 1854–1868,* Madrid, 1979.

11. Only the hesitation of Matías Romero, Juárez's representative, prevented Armento from traveling to Mexico. Guadalupe Appendini, "Cartas de Italianos dirigidas al Presidente Benito Juárez," *Excelsior 2* June 1998, 1 & 3.

12. For democratic catechisms, see Thomson (2002a: 190–201); Morales Muñoz (1990).

13. Not all Conservatives were as alarmed as this columnist of the Catholic newspaper *El Espíritu Público* (reprinted in *La Democracia* I, 52, March 2, 1864). The Conservative-Liberal general, Fernando Fernández de Córdova, who had commanded the Spanish army against Mazzini's Roman Republic in 1849 (an intimate companion at arms of General Ramón María Narváez, hammer of the Democrats during Spain's 1848) admitted in his memoirs that during the 1860s, ". . . in the presence of the grave events that developed in Europe, especially in Italy, I was persuaded that the march of ideas and the influence of opinion and of the various schools of liberalism could not be opposed by the resistance of governments; that those principles contained a more exact concept of justice and law, and that they had so invaded the consciousness of the people, that to oppose these currents was the equivalent to provoking ruins and tempests," Fernández de Córdova, *Mis Memorias* III, 452, cited in Jover (1971: 341).

14. The rustic *cortijos* of the family of the Moderado chieftain Ramón María Narvéz in Loja (Granada) were transformed during the 1850s into the "casas del campo" with gardens in the French style.

15. Callahan (1984); for a fuller development of the comparison between Mexico and Spain, see Thomson (2002a: 189–211).

16. Jean-Pierre Bastian, *Los Disidentes: sociedades protestantes y revolución en México, 1872–1910* Mexico, 1989; Juan B. Vilar, *Intolerancia y Libertad en la España Contemporánea. Los Orígenes del Protestantismo Español Actual* Madrid, 1984; and for a fuller development of this theme in Spain and Mexico, see Thomson (1998).

17. Archivo Municipal Histórico de Antequera, Orden Público 162, May 5, 1869, Alcalde, Antequera, to Civil Governor, Málaga.

18. For a fuller development of the case of Xochiapulco, see Thomson (1998).

19. For these labor practices during the 1960s, see Martínez Alier (1971).

20. For these local proto-indigenista patriotic histories, see Thomson (2002b).

References

Bayly, C. A. (2003). *The Birth of the Modern World 1780–1914.* Oxford: Blackwell.

Callahan, William J. (1984). "Was Spain Catholic?." *Revista Canadiense de Estudios Hispánicos* VIII: 159–82.

Deacon, Desley (1997). *Elsie Clews Parsons: Inventing Modern Life*. Chicago: Chicago University Press.

Dunkerley, James (2000). *Americana: The Americas in the World, around 1850*. London: Verso.

Esteban, Jorge de (1981). *Las constituciones de Madrid*. Madrid: Taurus.

Fallaw, Ben (2002). "The Life and Death of Felipa Poot: Women, Fiction, and Cardenismo in Postrevolutionary Mexico." *Hispanic American Historical Review* 82: 645–84.

García Cantú, Gastón (1974). *El socialismo en México*. Mexico: Era.

Godoy, Ricardo (1977). "Franz Boas and His Plans for an International School of American Archaeology and Technology in Mexico." *Journal of the History of the Behavioural Sciences* 13(July): 228–42.

Hale, Charles (1989). *The Transformation of Liberalism in Late Nineteenth Century Mexico*. Princeton: Princeton University Press.

Hewitt de Alcantara, Cynthia (1984). "Particularism, Marxism and Functionalism in Mexican Anthropology, 1920–50." In her *Anthropological Perspectives on Rural Mexico*. Boston: Routledge and Kegan Paul, pp. 8–41.

Hobsbawm, Eric J. (1988–98) *The Age of Revolution; The Age of Capital; The Age of Empire; The Age of Extremes*. London: Phoenix Press.

Jover, José María (1976). *Política, diplomacia y humanismo popular en la España del siglo XIX*. Madrid: Turner.

Knight, Alan (1994) "Popular Culture and the Revolutionary State in Mexico, 1910–1940." *Hispanic American Historical Review* 74: 393–44.

Landes, David (1998). *The Wealth and Poverty of Nations*. London: Abacus.

Levy Zumwalt, Rosemary (1992). *Wealth and Rebellion: Elsie Clews Parsons, Anthropologist and Folklorist*. Chicago: Chicago University Press.

Lewis, Oscar (1951). *Life in A Mexican Village: Tepoztlán Restudied*. Chicago: Chicago University Press.

Lida, Clara E. (2002). "The Revolutions of 1848 in the Hispanic World," in Guy Thomson (ed.), *The European Revolutions of 1848 and the Americas*. London: Institute of Latin American Studies, University of London.

Loveman, Brian (2001). *Chile. The Legacy of Hispanic Capitalism*, 3rd edn. Oxford: Oxford University Press.

Martin, Gerald (1989). *Journeys through the Labyrinth*. London: Verso.

Martínez Alier, Juan (1971). *Labourers and Landowners in Southern Spain*. London: Allen and Unwin.

Morales Muñoz, Manuel (1990). *Los catecismos en la España del siglo XIX*. Málaga: Universidad de Málaga.

Parsons, Elsie Clews (1930). *Mitla Town of the Souls and Other Zapoteco-Speaking Pueblos of Oaxaca Mexico*. Chicago: Chicago University Press.

Pérez Luzaró, Mariano (1853). *Historia de la Revolución de Italia de 1848 y 1849*. Madrid: no publisher stated.

Redfield, Robert (1930). *Tepoztlán. A Mexican Village*. Chicago: Chicago University Press.

Redfield, Robert (1934). *Chan Kom. A Maya Village.* Chicago: Chicago University Press.
———— (1940). "The Folk Society and Culture." *American Journal of Anthropology* 45(5): 731–42.
———— (1941). *The Folk Cultures of Yucatán.* Chicago: Chicago University Press.
———— (1947). "The Folk Society." *American Journal of Sociology* 56(4): 293–308.
———— (1950). *A Village that Chose Progress: Chan Kom Revisited.* Chicago: Chicago University Press.
Tenorio Trillo, Maurico (1999). "Stereophonic Scientific Modernism: Social Science between Mexico and the United States, 1880–1930." *Journal of American History* 86: 1146–87.
Thomson, Guy (1994). "The Ceremonial and Political Roles of Village Bands, 1846–1974," in William H. Beezley, Cheryl Martin, and William E. French (eds.), *Rituals of Rule, Rituals of Resistance. Public Celebrations and Popular Culture in Mexico.* Wilmington, DE: Scholarly Resources, pp. 307–34.
———— (1998). " 'La République au village' in Spain and Mexico, 1848–1888," in Hans-Joachim Konig and Marianne Wiesebron (eds.), *Nation Building in Nineteenth Century Latin America. Dilemmas and Conflicts.* Research School CNWS: Leiden, pp. 137–62.
———— (2001). "Garibaldi and the Legacy of Revolutions in 1848 in Southern Spain." *European History Quarterly* 31: 353–99.
———— (2002a). "Liberalism and Nation-Building in Mexico and Spain during the Nineteenth Century," in James Dunkerley (ed.), *Studies in the Formation of the Nation-State in Latin America.* London: Institute of Latin American Studies, pp. 190–201.
———— (2002b). "Memoria y memorias de la intervención europea en la Sierra de Puebla, 1868–1991," in Antonio Escobar Ohmstede, Romana Falcón, Raymond Buve (eds.), *Pueblos, comunidades y municipios frente a los proyectos modernizadores en América Latina, siglo XIX.* CEDLA: Amsterdam, pp. 145–68.
Thomson, Guy, with David LaFrance (1999). *Patriotism, Politics and Popular Liberalism in Nineteenth-Century Mexico. Juan Francisco Lucas and the Puebla Sierra.* Wilmington, DE: Scholarly Resources.
Vaughan, Mary Kay (1997). *Cultural Politics in Revolution. Teachers, Peasants, and Schools in Mexico, 1930–1940.* Tucson: University of Arizona Press.

Chapter 4

When Was Latin America Modern?
A Historian's Response

Alan Knight

Smart Oxford philosophy professors, seeking to find out which of their smart students were smartest, and perhaps worthy of becoming the next generation of smart philosophy professors, used to set examination questions along the lines of: "Is this the right question?"[1] The answers were hardly useful; they were never—to my knowledge—expanded into enlightening books entitled *Is This The Right Book?*; but they did serve an immediate purpose—sorting philosophical sheep from nonphilosophical goats. The question we are considering has something of the same instrumental quality: it has served to provoke some lively discussions and (unlike the previous example) it *has* been developed into a diverse and interesting book.[2] Pragmatically, therefore, the question works; but as a "heuristic device" for understanding Latin America, it is not much help. Like "Is this the right question?" it is laden with conceptual difficulties and, if we seek to go beyond such difficulties and "operationalize" the question in specific empirical contexts, it is difficult to make progress. In this, it resembles many other notional questions we might concoct: When was Latin America happy? When was Latin America good? And, the most obvious cognate question: When was Latin America traditional? On the other hand, it differs from other questions which, though they sound similar, are substantially different, since they are conceptually clearer and, to some degree, empirically operationizable, for example: When was Latin America literate? When was Latin America urban? When was Latin America industrial?

Even with these valid and useful questions there are several problems, of the kind that attach to all such sweeping interrogatives. Some problems can be quickly disposed of. First, the definition and derivation of

"Latin America" need not detain us. Whether there is an entity called "Latin America" that shares common characteristics, making it a valid unit of analysis (one of Huntington's building blocks of civilization, for example) (Huntington 1996), is not at issue; this is a separate question, worthy—perhaps—of another book: When or what was Latin America? We have enough on our hands as it is and can take "Latin America" to refer to the twenty republics conventionally defined as Latin American: the eighteen successor states of the Spanish Empire in the New World plus Brazil and Haiti.[3] If, for example, we found that Brazil was significantly different from the rest (perhaps it "modernized" earlier or later than Spanish America?), that finding would be of interest and we could incorporate it into our conclusions. But in my view it wasn't and it didn't, so the question doesn't arise. Meantime, we need not fret about what "Latin America" is; we have a very clear workable definition. Indeed, it's about the only clear workable definition in the whole discussion.

Second, any such question—when was Latin America modern, literate, urban, or industrial?—needs to be spatially disaggregated. Social and cultural change tends to be patchy and does not sweep across a huge landscape like a tidal wave. There are major differences between countries and even within countries. Chihuahua is not Chiapas and—with all due respect to President Kirchner—Santa Cruz is not the Provincia de Buenos Aires. Thus, even if the question is a broadly meaningful and manageable one—when was Latin America literate?[4]—we would probably conclude that, to be useful and convincing, the answer should be disaggregated by country, region, sector (e.g., city/countryside), sex, and age cohort/generation. And, since we are dealing with cumulative sociocultural processes (not sudden tidal waves), we would have to resort to broad chronological conclusions derived from time series: Mexico became literate in the 1940s, urban in the 1950s, and so on.

Furthermore, the bigger the unit, the more broadbrush is the answer. We can state more precisely and, I think, more usefully, when a region became urban or literate than when an entire country did. If Mexico (on average) became literate in the 1940s, Nuevo León had already crossed the threshold in the 1920s, while Chiapas did not do so until the 1960s (Wilkie 1970: 208–9). The same is true of countries within continents or even continents within the world. Every scalar increase brings, necessarily, an increased dispersion of values, thus, even if we could say when Latin America as a whole, on average, became modern, that would not tell us much about many component countries or regions within Latin America. Indeed, the conclusion might be positively misleading if it turned out that aggregate trends masked major differences; if, for example, while most of Latin America was urbanizing, some places were becoming more rural; or, to take a more plausible scenario, if industrialization in one area (like Monterrey) was offset by deindustrialization in another (such as the Bajío).[5] So, too, with modernity: maybe different bits of Latin America—the bits may be countries, regions, sectors, or subcultures—are moving in different directions: some are getting more "modern," some more "traditional" (or "antimodern"? or postmodern?). The assumption of common trends, therefore, depends upon and reinforces rather crude teleological notions: that entire societies are moving along a conveyor belt toward an ineluctable destination (modernity, industry, democracy).

There is a third standard problem to bear in mind. Anthropology has borrowed from linguistics the emic/etic distinction, which contrasts the concepts entertained by actors (anthropological or, in this case, historical) with those deployed by social scientists (anthropologists or historians) in order to understand those actors (Harris 1976). The distinction is sometimes hard to draw; but often it is obvious and crucial. We no longer explain disease in terms of divine punishment or mental illness in terms of diabolical possession. Regarding "modernity," we should distinguish—as some contributors in this book have—between the perceptions of actors and the analyses of social scientists (including historians). The question, therefore, bifurcates into: (a) when did Latin Americans—subjectively, emically—consider themselves and/or their society to be modern?;[6] and (b) when do we, as social scientists, consider that Latin America became—objectively, etically—"modern"? Sometimes— when subjective questions of identity are at stake—this bifurcation is very tricky. If the question was: "When was Latin America mestizo?" we would probably have to take the emic response as the basis for any etic conclusion—since to be "mestizo" is basically a question of subjective identity and if people consider themselves mestizos, there are good grounds for saying that they are.[7] The same would not be true of, say, "literate," since there are roughly objective criteria that can be used to measure literacy.[8] "Modern," to my mind, is more analogous to "literate," in that the term is regularly used as a supposedly etic description of individuals and societies; and scholars who use the term believe they are conveying some kind of objective information and are not merely relaying the subjective (emic) opinions of the actors themselves.[9]

The "emic" status of "modern" is much harder to establish. As Guy Thomson points out, it was not regular usage in mid-nineteenth-century Mexico, where "progress" and "civilization" were preferred (see Thomson, this volume). Around the same time, Sarmiento proposed his famous dichotomy between "civilization" and "barbarism"; on the few occasions when he used "modern," it seems to serve as a general description much like "contemporary" (thus, barbarism is regarded as part of the "modern world") (Sarmiento 1961: 38). Scholars have referred to Sarmiento's "ideas on nation building and modernity" and (perhaps) his "emerging visions of modernity" (Viñas 1994: 214–15);[10] but this is a slightly risky business, since it imputes to Sarmiento concepts that he—an articulate, well-read intellectual—did not explicitly and regularly use. We, therefore, have to take on trust two major imponderables: the scholar's own understanding/ definition of "modernity" and the assumption that Sarmiento, largely implicitly, shared that understanding/definition. This calls for a lot of trust.

From personal knowledge, I do not think "modernity" was common currency in the debates surrounding the Mexican Revolution (1910–40): Mexicans argued about capitalism, socialism, liberalism, progress, civilization, and democracy—but rarely "modernity." At the local level, communities were sometimes divided between those supposedly eager for "progress" (correctos) and those who resisted (tontos) (Redfield: 1930). More recently, it seems, this old dichotomy has been reformulated to embrace "modernity": in a contemporary Yucateco town, Pustunich, people talk about being "modern," which they contrast with "traditional"—the latter connoting "poor," "humble," "peasant," and Maya (Greene 2001: 418). In doing so, they seem to be following social-scientific precedent: James Wilkie's well-known but contentious "poverty index," devised in the 1960s, took seven indicators of "non-modern standards of living" that were equated with poverty; they included illiteracy, speaking an Indian language, going barefoot or wearing sandals, and eating tortillas (Wilkie 1970: 205 ff.). These roughly correspond to the characteristics of the *tontos* of Tepoztlán in the 1920s and the "traditionals" of Pustunich in the 1990s.

I am not aware of any systematic study of what popular cultural notions of modernity—and tradition—currently prevail in Mexico, still less in Latin America as a whole. If, however, we narrow the focus to elite political notions (which are easier to get at, since elite *políticos* talk a lot and get into print), a tentative hypothesis might be advanced. My sense—heavily based on Mexican evidence—is that "modernity" sneaked in under the radar screen in fairly recent years, as a kind of etic invader. It represented an extension into everyday (emic) thought and parlance of the supposedly etic concepts "modern," "modernity," and "modernization," concepts coined by social scientists in the later twentieth century and now espoused, mainly, by neoliberal technocratic *políticos*, their intellectual sidekicks, and journalistic apologists. The success of this invasion was, I think, quite limited: "modernity" has not become a basic conceptual weapon in the armory of recent Latin America politics (in the way that "democracy" and "democratization" have, for example). Still less, of course, has the standard antithesis of modernity—tradition—assumed such a role. I cannot think of a single Latin American party of the last 100 years that boasted either "moderno" or "tradicional" in its official label.[11] Where "modern" and "modernity" do figure is in more recent political—and, at Pustunich, cultural—discourse.

There is a possible comparison here with "populism." The concept of "populism" was developed as an academic, political science tool, supposedly to explain certain regimes and movements of the mid-twentieth century. Those regimes and movements, to my knowledge, rarely if ever called themselves "populist"; for many, the term would have been alien, even unknown.[12] Populism arose as an academic, etic notion. Much later, after the classic populist movements and regimes had disappeared, the term acquired wider currency, probably thanks to the efforts of a later generation of intellectuals, *políticos* and opinion-mongers, who found it a useful term of partisan politics—thus, of emic discourse. Again, President Salinas and his neoliberal cohort would be the prime Mexican culprits. By the 1980s and 1990s, about the time that "modern" also became a positive (emic) buzz-word, it was common to hear off-the-cuff denunciations of political and economic "populism" (Knight 1998: 226, 241, 243–44). In both cases, I suggest, a supposedly "etic" term, generated by social scien-tists and intellectuals, was "emicized"—it entered political debate as a partisan term. These discursive trends are, of course, interesting and in some instances important. However, both "modern" and "populist" are problematic emic terms, because they tend to be conceptual one-way streets (the biggest one-way street in today's political lexicon, is, of course, "terrorist"). "Populists," both past and present, did not consider themselves "populist"; it was a term invented by social scien-tists and then appropriated by a later generation of *políticos*, who turned it into a political slur. "Modern" also had—and has—etic claims to social-scientific status (which I shall address later), but its "emic" status is similarly murky. In the nineteenth and early twentieth century, as I have said, it was rarely used; and when, more recently, it caught on, it too became a partisan term, in this case, a positive one. "Four legs good, two legs bad," chanted the beasts of Animal Farm; "modernity good, populism bad" goes today's official chorus. This is—shifting the metaphor—a one-way street because, to my knowledge, few or no political actors proudly declare themselves either populists or enemies of modernity. Populism is—in emic political discourse—a vague insult, modernity a vague universal good to be applauded, like peace, prosperity, motherhood, and apple pie. Thus, whereas it would be quite legitimate to consider both the emic and etic import of, say, "Marxism-Leninism" (it is meaningful to ask—albeit difficult to answer—the question: is Castro really a Marxist-Leninist and, if so, when did he become one?), the same cannot be said of "modern." It is not a very useful emic concept.

But is it a useful *etic* concept—useful, that is, to social scientists who seek to understand Latin America, past and present, irrespective of the opinions of historical actors? Here, two basic questions arise: the *conceptual* question of what the term means; and the *empirical* question of how—excuse the jargon—it is to be "operationalized." By "operationalization" I mean the following: How is the concept to be applied to a messy reality in such a way that useful and cogent conclusions emerge? My formulation is based on the pragmatic idea that the value of a concept derives from its usefulness—not, in contrast, from its membership of a privileged pantheon of Platonic essences. What we should ask of big concepts is not what company they keep or which genius invented them, but, in the words of Janet Jackson, "What have you done for me lately?" As for "operationalization" (and I apologize for this octosyllabic monster), it requires that, once a reasonably clear "organizing concept" is in play, we can find reliable and relevant empirical data that it usefully organizes. There is little point in having an impeccable concept that defies all empirical research.[13]

Regarding definition, it is a matter of concern that "modern" has so many diverse connotations. It is reminiscent of Barber's apt summation of the concept of "totalitarianism": "A conceptual harlot of uncertain parentage, belonging to no-one but at the service of all" (B. R. Barber, quoted in Giddens 1987: 296). For, as I have said, one criterion of a good (etic) definition or concept is that it should be clear. Yet "modern" has multiple meanings. For some—including, it seems, Sarmiento—it simply means "recent": clipper ships were "modern" until they were superseded by steamships (Inkeles and Smith 1974: 292). This is harmless—and, in a modest way, useful—but it is irrelevant to our discussion. According to conventional historical usage, the "modern" history of Europe conventionally starts ca. 1500 (or whenever the Middle Ages ended and medieval gave way to modern) and it is conventionally divided into "early" modern (ca. 1500–1789 [?]) and (late?) modern (post-1789[?]).[14] Works of history—like the monumental *Historia moderna de México*—adopt this conventional usage (Cosío Villegas 1955–65). There is nothing wrong with this, but it is simply a way of flagging a time period without using numbers. In similar fashion we use "Classic" to denote Mesoamerican history ca. 100–ca. 800 without necessarily implying distinctively "classic" features (Knight 2002: 26). Over time, of course, such usage may become dated. In art, architecture and technology (among other things) what was modern soon becomes not-modern—traditional, passé, out of date, admired by Prince Charles. Clipper ships are no longer "modern." Historiography often moves at a more glacial pace, but the historians of the distant future may one day decide to revise conventional nomenclature and rethink, even rename, this ever-lengthening "modern" period (ca. 1500–>?). So long as the convention prevails, our question is easily but uselessly answered: Latin American became modern, like everywhere else, ca. 1500 (we might as well say 1492); so, Latin America was born modern. It was never anything else.[15] Plenty of scholars take this view; some even believe there is a profound rationale to it. "Modernity," they argue, was somehow spawned by the European encounter with the "New World": "America was the first periphery of modern Europe."[16] This is both factually wrong (there were several earlier peripheries, as both the Crusaders and the Teutonic Knights were well aware) and also conceptually confusing, since it seems to take the ancient phenomenon of frontier expansion as somehow diagnostic of "modernity."

Of course, scholars like to think that their categories—in this case, chronological categories, like "modern"—have intrinsic meaning and are not just verbal substitutes for numbers (like 1492 or 1500). Most, therefore, assume that "modern" is more than a neutral chronological label and that it carries certain specific qualities, which are characteristic of "modernity." The modern period is modern precisely because it introduced and favored these qualities. But what are they? Laurence Whitehead (this volume) places great stress on the alien or imported nature of "modernity." Now, whether "modernity"—however we wish to define it—came to Latin America from outside or was domestically nurtured is an empirical question, and possibly one of real interest. But it has nothing, that I can see, to do with the *definition* of modernity (unless, that is, we want to rewrite the English dictionary). Modernity *may* have come to the Americas from Europe, like cattle and coffee, steel and smallpox, but that does not warrant it carrying a permanent "made in Europe" stamp. Tridentine Catholicism came to the Americas from Europe: was that modern? It was, in its day, chronologically modern, since it took little time to leap the Atlantic following the Council of Trent; but its beliefs and practices do not square with—indeed, they are antithetical to—most social-scientific criteria of "modernity."[17] More recently, Latin America has "imported" a great many North American ideas, policies, and commodities. Some might qualify as "modern" by those criteria, but the mere *fact* of being imports tells us nothing, except that they are imports. Evangelical Protestantism is not, by many criteria, notably "modern." Furthermore, such a viewpoint, stressing Latin America's chronic importation of modernity, encourages an oddly diffusionist notion of cultural change (with modernity flowing unidirectionally from center to periphery); it neglects multiple invention and discovery (the Native Americans had hit upon art, astronomy, agriculture, religion, urbanization, and state building without requiring outside instruction); and it appears to deny Latin America an autonomous capacity to generate its own "modernity." It also begs the question of European—or North American—"modernity." Where did it come from? Why does the center appear to enjoy a monopoly when it comes to making modernity? Indeed, can we even trace cultural flows in this mercantilist fashion, with the center enjoying a permanent trade balance, and each cargo being tallied and labeled as it traverses the Atlantic from east to west?

If, as I have said, a serious problem derives from the sheer slipperiness of the notion of "modernity," it should be recognized that some social scientists have made serious attempts to clarify and deploy the concept. Indeed, for a time "modernization theory" was a powerful paradigm that bestrode Latin American studies. Thus, while "modernity," as an "organizing concept" lacks the intellectual lineage and related canon of, say, "capitalism" or "democracy," it does have its classic texts, thinkers, and postulates. It would be a mind-boggling— and, frankly, rather mind-numbing—task to summarize these. I shall simply list some of the notions or postulates that have been taken as diagnostic of "modernity" and the process of "modernization." Modernity implies: rationality and rationalization; secularism; "disenchantment"; literacy; urbanization; "achievement-orientation" (as against ascription); and bureaucracy.[18] It involves (institutionally) mass education, industrialization, rapid transportation and communication.[19] It confers (psychologically) access to information, openness to new ideas, awareness of time, a commitment to long-term planning, and an appreciation of technical knowledge.[20] It is associated (politically) with informed participant citizens, a sense of personal efficacy, freedom from traditional ties, and flexible cognition (Inkeles and Smith (1974: 290–4). It brings capitalism, computers, and fax machines

(Renegger 1995: 42). The list could go on: promiscuity without responsibility, we might say, is the prerogative of "harlot concepts" throughout the ages.

We could no doubt pare down the concept somewhat, noting what are central criteria and what are secondary.[21] But the sheer plethora of criteria, and the endless permutations that they produce, raise serious problems, since different criteria will, of course, yield different conclusions when we proceed to empirical "operationalization." Not all big ideas are quite like this. Democracy—liberal, representative, "polyarchic" democracy—is reasonably well defined, thus is amenable to arguments concerning its causes, limitations, social correlates, and putative "consolidation." The definition of capitalism is open to debate: but the debate revolves around two basic approaches, the "circulationist" and "productionist," the latter being more parsimonious and rigorous. Given the relative simplicity of the debate, it is not difficult—when dealing with particular cases in the pragmatic way I have advocated—to use both approaches, if so desired. When it comes to "modern" and "modernity," however, the broad range of possible criteria makes such an even-handed Salomonic solution impossible; and the kind of definitional consensus that facilitates debate about democracy is conspicuously lacking. Instead, the variables are slithering all over the place like beads of liquid mercury.

Given such a plethora of criteria, it is pertinent to ask whether they correlate—or, at least, display a mutual "elective affinity"—and whether, therefore, they merit a common catch-all label, "modern." It may be quite sensible to talk of urbanization, industrialization, bureaucratization, secularization, and any number of other "-izations," but is there any good reason to assume that they hang together in a historically and/or logically coherent syndrome? Modernization theorists believe, as they have to, that there is such a syndrome (or "ethos" or "mentality") (Inkeles and Smith 1974: 16, 291). For, without some genuine and meaningful relationship between the various—and disputed—attributes of modernity, these become no more than individual traits, randomly recombinant, lacking any systemic cohesion. "Modernity" would then become useless as a higher-order category of analysis, even though we could productively continue with many of its supposed constitutive elements (industrialization, bureaucratization, secularization, etc.). So, does such a "modernity" syndrome exist? I would argue that such a syndrome can be discerned—and, therefore, the idea is not a total chimera—but that it cannot be deployed as a grand cross-national, cross-temporal explanatory model. It is possible to talk—cautiously and historically—of "modernity," but that is a far cry from believing in the grand delusions of modernization theory.

The most convincing description of modernity as a syndrome—as a bunch of genuinely correlated and "electively affine" characteristics—is that given by historians of the European Enlightenment and of the "New Philosophy" that it introduced. Working on a specific time and place (Europe, ca. 1650–1800), they discern a set of ideas that hang together both logically and historically (that is, they tend to be held by the same people and repudiated by another group of people) and that have a decisive impact (thus, we see a sharp contrast between "before" and "after" the Enlightenment). The cornerstone of these ideas was rationalism, not in the (Weberian) sense of ends/means rationality, but rather the principle of subjecting all knowledge and practice to the bright light of reason.[22] Thus, revealed religion, faith, magic, superstition, confessional politics, and divine right were all called into question; the ideational pillars of the old regime (broadly defined) were systematically undermined; ancient institutions—Church, Crown, Inquisition, aristocracy—lost their legitimacy. This—as some philosophers concur and even lament—was a major intellectual shift, a Copernican "revolution," if you like, when a materialist and humanist universe replaced a God-centered one.[23] It happened fairly rapidly and affected the whole of Europe (Israel 2001: v–vi, 24, and passim). Naturally, it provoked a powerful reaction, a "Counter-Enlightenment," which involved both argument and repression.

This historical story makes sense because the empirical evidence is strong and the story possesses a degree of internal cohesion and cogency (we can see why an attack on revealed religion, justified by reason, would lead to a critique of Church and Crown, and, potentially, of all sacralized and prescriptive authority). Of course, the story may be wrong (I am not an expert on this period or problem); and it may certainly exaggerate the social impact of these ideas (Keynes's famous dictum concerning the power of ideas strikes me as one of sillier things he said).[24] Nevertheless, if we confine our analysis to the realm of European elite thinking in the period, the fact of an intellectual sea change seems clear, and this sea change could be summed up as the birth of modernity.[25] The choice of catch-all label derives, in part, from the people themselves, who espoused the cause of the "moderns" against the "ancients" (so it has some, perhaps flimsy, "emic" validity) (Jones 2002: 188–9); but, more important, it is historically a reasonable label to use, since it captures the idea of something new, a decisive break with the past, and a repudiation of tradition (i.e., the inherited ideas and institutions of the old regime).

Enlightenment "modernity," therefore, corresponds to a "mood" and contributes to a—critical, rational, scientific, "disenchanted"—mentality.[26] While it depended on certain institutions and forms of "sociability" (cafes, salons, masonic lodges, corresponding societies) and, of course, it had a broad social and political as well as cultural impact,[27] it cannot be defined or explained in terms of social organization: big cities and (merchant) capitalism preceded the Enlightenment; industrialization and mass education came a good deal later. "Modernity" is, therefore, a philosophical or intellectual creature, but its birth is a historical fact, with a time and place.

The European Enlightenment was, therefore, a one-off. It had a broader impact: indeed, Latin America was an eager recipient of Enlightenment ideas (here, Whitehead's diffusionist model works, to an extent).[28] But, just as the impact of these ideas in Europe may be debated (how broad, deep, and enduring were they?),[29] their echoes around the world become, with time, increasingly confused and chaotic. Like the Big Bang that started the Universe, the European Enlightenment leaves behind it a complicated cosmos, with bits flying apart, lumpy galaxies scattered across vast voids, and a pervasive background hiss reminding us of the distant point of origin. Thus, while we could trace some of the supposed characteristics of modernity back to that point of origin, dispersal and variegation have since proceeded apace; furthermore, *pace* Keynes, we should resist the notion that major social and political changes—urbanization, industrialization, state building, bureaucratization—obey intellectual imperatives, when, in fact, they more likely derive from the interplay (dialectic?) of political and economic interests (classes?), with intellectual explanations/justifications often being invoked *ex post facto*. So, the subsequent story of Enlightenment ideas is one of selective appropriation, distortion, and repudiation.[30] Some ideas have survived and prospered (consider, e.g., James Scott's "high modernist ideology") (Scott 1998: 4 ff.); some have made limited progress (humanism and secularization); some continue to be fought over (citizenship and the rule of law). The story is not one of linear advance or retreat, but vicissitudes over time and place. Applied science and technology have flourished (they bring discernible benefits, at least to some); but secularism has faced mixed fortunes, while revealed religion, although it eventually declined in the ancestral home of the Enlightenment, Europe, has flourished elsewhere, not least in the United States (which was, more than most countries, a child of the Enlightenment). Latin America has become more religiously diverse, but there has been no triumph of secularism; indeed, in recent years, a revivalist Pentecostalism has gained ground on a more bureaucratic, restrained, perhaps "rational" Catholicism.[31] In the realm of social organization, bureaucracy has grown (as, until quite recently, states have also grown); but whether—in Latin America or elsewhere—bureaucracy follows rational, meritocratic, Weberian principles or, rather, betrays clientelism, nepotism, and *empleomanía*, is open to question. Industrialization, today, is a Third World (especially Chinese) phenomenon; greater São Paulo is more industrial (*ergo* more "modern"?) than Chicago. Enduring inequalities cast doubt on the degree of social mobility; some would even argue that the supposedly rational free-market capitalism espoused by recent policy makers is itself an irrational quasi-magical belief; or, more plausibly, it is a rhetorical justification for regressive, capital-friendly policies (sorry, Keynes, wrong again). In each of these instances, we may choose to reify ideas and see enlightened "modern" principles of rationality and secularism underpinning, say, bureaucracy or mass education or industrialization.[32] But in reality these principles are regularly undermined, resisted, and distorted, while the social trends that they supposedly underpin are the product less of *ideas* than of *interests*, both political and economic. Power and wealth count for a lot more than the imperatives of "modernity." Perhaps the clearest examples are provided by those episodes when "Enlightenment" projects were seriously and systematically attempted, for example, with Mexico's "socialist education" (and related secularizing policies) in the 1930s: the state could exhort, build schools, and repress dissent, but creating a "new man"—and a "new woman" and "new child"—steeped in Enlightenment values proved extraordinarily difficult (Knight 1994). This does not make Latin America particularly unusual. From the Scopes trial to contemporary Creationism, the United States often seems determined to repudiate its rational Enlightenment inheritance.[33] A half century of Marxism—and very authoritarian Marxism—has not extirpated traditional religion, folk medicine, and magic in China. And, needless to say, the history of Europe in the 1930s and 1940s is hardly an advert for enduring Enlightenment values in the birthplace of the *Aufklärung*.[34]

What this tells us, I think, is that, even if there is a discernible "dawn of modernity," associated with the European Enlightenment, the subsequent implementation of "modern" ideas has been patchy and partial (even in Europe). It has not been a linear story of advance, but rather a crabwise dialectic of advance, pause, and retreat, the final itinerary being determined more by the power of interests than of ideas. Some Enlightenment notions—applied science, for example—have proved successful, not least because they respond to perceived needs (for power and productivity). Others, such as secularism, have experienced mixed fortunes. Implicit in this topsy-turvy narrative is the lesson that modern values do not form a coherent whole; they are not bound together by unbreakable bonds of elective, still less essential, affinity. Bureaucratic and technologically sophisticated regimes (e.g., Nazism and Stalinism) do not conform to "modern" notions of reason and freedom. The contemporary United States is highly "modern" in some respects, recalcitrantly antimodern (traditional?) in others (hence the power of faith, religion, astrology, Creationism, new age cults, and the like). Sometimes these contrasting attributes are to be found in different sectors and locations (hence, the "red" heartland and "blue" seaboards of the United States); but often they jostle for supremacy within individuals.

Thus, even if the modern/Enlightenment syndrome once— ca. 1750—constituted a coherent intellectual syndrome, a set of logically and historically related beliefs, it has, over time, lost its coherence. The Big Bang has given way to a lumpy, disparate, expanding universe. We may consider more specific historical analogies. Both the Norman Conquest of England in 1066 and the Spanish Conquest of Mexico and Peru in the early sixteenth century represented dramatic moments when history—the history of England, Mexico, and Peru—changed decisively and permanently. At the moment of conquest it was possible to define the "contributions" of the conquerors: in the first case, Norman French, feudalism, and motte-and-bailey castles, in the second, Catholicism, coinage, steel, smallpox, and sheep (among other things). We could call these the Norman and Spanish Conquistador complexes. Before long, however, these new elements fused with "native" beliefs and practices. Norman French blended with Anglo-Saxon to produce Chaucer's English. Mexican Indians became sheep-farmers and their religion developed as a syncretic blend of old and new beliefs. When, centuries later, scholars sought to separate the cultures of conqueror and conquered, they often found it difficult (e.g., Clews Parsons 1936). So, while it was meaningful to talk of Spanish (Castilian, Andalusian, Extremeño) or Mesoamerican (Nahua, Maya, Totonac) cultures in the 1520s, it makes much less sense 400 years later. Similarly, the fact that we can perhaps discern a form of Enlightenment modernity in Europe ca. 1750 does not mean that this historically specific syndrome can or should serve as a benchmark for all subsequent societies. It did usher in new ways of thinking, which have spread to the rest of the world, Latin America included. But as it spread it was selectively appropriated, utilized, and refashioned (as well as ignored and repudiated). What remained was no longer a coherent syndrome, but rather a set of shifting attributes whose "elective affinity" was often lost. Religious organizations—the Vatican, the Company of Jesus, Opus Dei—readily espoused technology, the mass media, and bureaucratization. Secular authorities fell prey to irrationalism, personalism, and despotism. The massive cities of the "modern" period were not havens of sociability and *doux commerce*. Countries that once prided themselves on their "European" modernity and development—Argentina, Chile Uruguay—pioneered the barbaric political authoritarianism of the later twentieth century.[35]

In short, the comforting certitudes of modernization theory proved disappointing. According to some critics, they were no more than the ethnocentric extrapolations of North American social science; or, worse, the insidious propaganda of North American government and big business. This was somewhat unfair: modernization theory, though a poor guide to history and politics, had some respectable ancestry (Weber, Toennies, Maine) and was not devoid of insight. But if it was to be of any use, it had to be unpacked and deployed critically and selectively. As a *passe partout* to historical understanding—or political action—it was greatly overrated. By ca. 1980—if not before—this sobering conclusion seemed almost consensual.[36]

Yet in recent years modernization theory—or approaches that, under different names, persist in using "modernity" as their central "organizing concept"—have mysteriously revived. In part this revival reflects the neoliberal project that arose from the ashes of the debt crisis, committed to building a "modern" free-market Latin America in the (supposed) image of the United States. This—"southern"—commitment to "modernization" obeyed political and economic motives, some of them—as usual—rather more self-interested than their high-minded rhetoric allowed. Again, ideas (of "modernity") masked interests (class and sectoral). Neoliberal projects, like Salinismo, trumpeted their modernity and, as I suggested earlier, castigated the purblind "populism" of their opponents. But neoliberalism in the south also coincided with a "northern," intellectual and academic reaffirmation of "modernity" as a central explanatory concept.

Here, I think, two very different currents of thought intersected. First, U.S. political science underwent a partial "cultural turn." Some political scientists reacted against rational choice and number crunching; some were influenced by the "culture wars," which they perceived in the United States and elsewhere; and some—the older generation—just refurbished the culturalist-cum-modernization theory they had been peddling for years, thus proving that, if you carry on long enough, fashion comes full circle and old hat becomes *la nouvelle vague.*[37] Second, a (usually) younger and more radical group of scholars, trained in or influenced by literary and cultural studies rather than the social sciences, also placed "modernity" at the center of their often confusing "analyses." (To call modernity their "organizing concept" would be misleading, since conceptual organization is not their strong suit; indeed, they often make a virtue of imprecision and intellectual skittishness.) These scholars probably knew little about modernization theory and certainly would have been horrified to be linked to the likes of Parsons and Rostow. In contrast to the latter, they often regarded modernity and modernization as ambiguous or even downright bad. Many, having read their Foucault et al., conceived of the Enlightenment as a thoroughly Bad Thing.[38] But, even if their normative assumptions were very different, they still peppered their writings with confident references to modernity. What is more, while the older, social-scientific generation at least attempted definitions of modernity/ modernization, the younger generation tended to take it as a given; indeed, they often have "modernity" enter like some pantomine villain—stage-right, swathed in black, to the hisses of the audience.

In this recent work, modernity is often contrasted not with its old alter ego "tradition" (a concept that now elicits greater and justifiable skepticism) but rather "postmodernity." This new dichotomy generates huge problems. First, the contrasting criteria of "modernity" and "postmodernity" are vague. Second, criteria aside, there is serious disagreement concerning historical pedigree. If modernity is associated with the Enlightenment project (a proposition I have cautiously endorsed), then are nineteenth-century reactions against the Enlightenment (Romanticism and later modernism) post- or anti-modern? Or is postmodernism a product of the twentieth century, a successor to and reaction against *modernism?* (Hollinger 2001: 10–4). The confusion arises, I think, because labels and categories have been casually transposed across disciplines. Modernism and postmodernism were originally terms used to describe trends in art, architecture, and literature. No doubt they made some sense; at least plenty of intelligent people used them and they seem to have facilitated debate, which is what such terms are meant to do (Nicholls 1995; Fascina and Harrison 1982). But to transpose artistic, architectural and literary categories to political, historical, and socioeconomic analysis is a risky business or, to put it more strongly, crass intellectual imperialism. It does not make sense to seek our basic historical or sociopolitical concepts in the work of Baudelaire, fine poet though he was.[39] Nor should we slap such "cultural" labels on entire historical periods and societies: to call Latin America quintessentially "baroque," for example, combines both a basic category error and a gross reification.[40] "Baroque"—like "modernist"—may serve as a useful shorthand description of, say, an architectural style, but to talk of "baroque politics" or "baroque economics" is unhelpful. Could we speak of mannerist mercantilism? Did Viceroy Revillagigedo the Elder practise rococo politics? I am not denying that there are demonstrable connections between, say, architecture and politics; but these connections do not justify the transposition of architectural–or artistic or literary— categories to entire historical periods or broad sociopolitical trends. We might call this bizarre approach the Harry Lime theory of history. Harry Lime, of course, believed that the cultural creativity of the Italian Renaissance was indissolubly bound up with the murder and mayhem of the period; in contrast, Switzerland "had 500 years of democracy and peace, and what did that produce? The cuckoo clock."[41] In other words, historical periods have a seamless unity or essence. Art, politics, and economics conform to a single logic and march together in lockstep. It is, therefore, legitimate to lift, say, artistic labels and slap them on entire periods and societies. Thus we get postmodernism—as a historical or societal, not literary, label. But, *pace* Harry Lime, history is not like that. Though there are certainly connections (between art, architecture, government, markets, and so on), there is also enormous slippage. And some human activities are simply not commensurable. If "cultural" terms can colonize history, politics, and economics, why not the reverse? "Capitalist" art is now a hollow cliché; but why not, as mentioned earlier, mannerist economics or rococo politics? If we were in a ludic mood we could no doubt come up with plenty of such conceptual crossbreeds: Ricardian novels; neo-Keynesian canvases; marginalist musical comedies. But all would be like bad-tempered mules, impossible to work with and incapable of producing useful progeny.

To conclude. The notion of "modernity" embodied in late twentieth-century modernization theory, though at least it was occasionally explicit, is not historically much use, since it lumps and reifies excessively, assuming a "syndrome" of ideas and practices which, in practice, often diverge and display scant "elective affinity" (even in the United States). Furthermore, while particular versions of modernization theory may be explicit, the genre as a whole is too diverse and eclectic; dozens of variables are promiscuously procreated and then irresponsibly turned loose, like maladjusted teenagers, on a vulnerable public. In its more egregious form, too, modernization theory became an apologetic for North American society, government, and foreign policy. Thus, it was rightly criticized for its ethnocentrism, teleology, and intellectual imperialism. Telling criticism may have blunted its academic appeal, but in the "real world," especially the real world of U.S. policy, similar ideas—ethnocentric, teleological, intellectually imperialist—are still influential.[42] When it comes to understanding history in general, therefore, modernization theory is not a lot of use: as a theory it is too vague and slippery; and actual societies do not seem to behave in the way modernization theory would have us believe. As a descriptive concept it is inferior to, say, "democracy" (which at least can be reasonably precisely and consensually defined) and as an explanatory concept it is inferior to, say, "capitalism" (which, while it may not connote mechanical "laws of motion," does at least capture some basic features of a given economic system).

The recent academic revival of "modernity" as a supposedly useful organizing concept also serves little practical purpose. The discovery that "culture matters" is a reinvention of the wheel; and, often, a rather clunking, inefficient wheel (Harrison and Huntington 2000). More noticeably, "modernity" (and its trendy cognates, modernism, postmodernism, even post-postmodernism) represent—for serious history—a hostile takeover by the asset-strippers of lit crit and cultural studies. Concepts that may have been useful in their domestic domains can become noxious nuisances when they are transported to quite different environments: like the Castilian sheep introduced to Mexico, they proliferate and play havoc with the local ecology (Melville 1994).

For these reasons it is very difficult to answer—even to start to answer—the question, "When was Latin America modern?". (Here I find myself in general agreement with Peter Wade, this volume.) We could certainly disaggregate the great catch-all of "modernity" or "modernization" and pursue individual variables; we could, if we had plenty of time and rather limited imagination, take Kahl's 7 scales of modernism or Inkeles and Smith's 24 variables and try to extend these scholars' synchronic analyses over time, backward and forward. But, if we attempted such operationalization, we would soon encounter the problems I mentioned at the outset: the data would usually not be available (for the past); the answers would depend on the unit of analysis (continent, country, region, city, locality, sector, age group, sex, class, ethnic group); and the conclusions would necessarily be vague ("Mexico became a majority literate society in the 1940s"). Quite what literacy meant, how it was used, and what cultural or political impact it had, would all remain moot. In other words, many of the interesting questions would remain unanswered. And, at the end of this exercise, I do not believe we would have identified a coherent "modern" syndrome: we would simply have plotted several different variables (literacy, urbanization, industrialization), whose relationships are themselves highly variable and not, I would say, constitutive of a distinctive "modern" identity, which can be usefully contrasted with a "traditional" or "non-modern" or "premodern" or even "postmodern" identity.

I arrive at the pernickety conclusion—though a conclusion that some even more pernickety historians might query—that there is just one specific, discernible, historical "modernity," which is the syndrome of ideas associated with the Enlightenment: rationalism, secularism, humanism, materialism. These were new, they were seen to be new, and they displayed a common logic, an "elective affinity." The most obvious were located in the realms of science, "natural philosophy," and history; it would be risky and even wrong to transpose them to, say, music or poetry. Though they originated in Europe, they soon spread, not least to Latin America, where they were often enthusiastically received. In some crude sense, therefore, we could say that modernity, in the form described, reached Latin America in the eighteenth century; and, though its carriers were usually outsiders, like Humboldt, it soon acquired its eager Latin American exponents (Whitaker 1961; MacLachlan and Rodríguez O. 1980: 288–91).

But in the Americas, as in Europe, the incidence and impact of these new ideas were highly variable over time and place. Vast swathes of terrain remained untouched, or were policed by anti-Enlightenment, antimodern authorities. Europe had its Holy Alliance, Latin America its Ultramontane bishops and know-nothing caudillos. So, the outcome was patchy. Spain's Bourbon reformers were keen for economic development; they promoted mining and military technology; they sponsored scientific expeditions; they frowned on the "superstitious" excesses of popular religion; and they sought to count and control their wayward American subjects. Some of this was arguably "modern." But, for obvious reasons, they did not favor "modern" notions of individual freedom, self-determination, and equality. The wars of independence, therefore, were not simple struggles of tradition against modernity, but rather conflicts between heterogeneous rival coalitions, each of which contained both "modern" and "traditional" elements.[43] After independence, as recent research has stressed, "modern" notions—citizenship, representative democracy, elections, civic association, public debate, print media, equality before the law—took root and, in some pockets of Latin American society, flourished (Forment 2003; Annino 1995; Guerra, Lempérière, et al.

1998; Guardino 2005). Yet, as Sarmiento and others lamented, barbarism and old regime obscurantism remained powerful. Aspects of modernity were apparent; but had modernity reached critical mass, such that Latin America could be said to be "modern"? Again, the question is unanswerable, since the terms of reference are too vague and the data are inadequate.

More realistically, we can say that, like the matter that spewed forth from the Big Bang, the original Enlightenment soon began to fragment, diffuse, and lose its initial coherence. Entropy increased; things got increasingly complicated and disordered. The success, over time, of "modern" ideas depended, in great measure, on their demonstrable truth and—more important—their perceived utility. "Truth" counted in respect of applied science and medicine, so these fields progressed (and, indeed, Latin Americans made original contributions, and were not merely consumers of "First World" expertise) (e.g., Coutinho 2003: 76–100). Yet, alongside modern medicine, *curanderos, brujos*, and shamans also flourished. Governments and armies eagerly took advantage of the benefits of applied science and technology (railways, telegraphs, radio, machine guns, helicopter-gunships), the better to control, monitor, tax, and repress their populations. In this respect, regimes that spurned Enlightenment notions of representation and civility were often keenest to stock the technological armory of the "national security state." The accoutrements of "civilization" could serve the interests of "barbarism" (Sarmiento would have been shocked, but no inhabitant of post-1918 Europe should have been surprised). Cities and industry grew, but—for better or worse—these trends did not herald a secular, humanist, tolerant, rational society.

In short, the historically identifiable "modern" syndrome of the Enlightenment was—*sub specie aeternitatis*—quite short-lived. The dream of the *philosophes*, the waking fear of obscurantists and reactionaries, was never wholly realized. The syndrome fragmented and fell prey to particular interests (the fate of most great ideational systems, from Christianity to Marxism). Bits of modernity survived, like galaxies in the void of space, or Norman French sedimented in the basic vocabulary of modern English ("beef," "castle," "uncle"). Some of the original ideas of Enlightenment modernity, though at best partially fulfilled, have retained their appeal and aspirational value ("on ne tue point les idées" ["ideas cannot be put to death"], as Sarmiento defiantly scrawled on a wall in Mendoza, before departing for exile in Chile in 1840) (Sarmiento 1961: 9). Citizenship, equality before the law, and free expression, though often infringed or denied, are today indelible features of Latin America that trace back to Enlightenment origins. On these grounds, Latin America is more modern now than it was in, say, 1600 or even 1700. But to attempt greater precision—to determine *when* Latin America became modern—is to tackle an impossible question, since it is a question that marries a nebulous concept (modernity) with intractable or inaccessible data. So, to return to smart philosophers, we might remember the advice of one of the smartest: "Whereof one cannot speak, thereof one must be silent."[44]

Notes

1. I use the past tense because the example is over 30 years old. While I cannot speak for philosophy examination questions today, I cannot see today's Oxford historians setting such quirky questions, for fear of students complaining, litigating, or getting 2/2s.
2. I attended the entire conference and heard (in some cases also read) the papers. However, this essay is not meant to be a comprehensive survey, but rather a personal response to the question and some of the various answers it elicited.
3. Haiti is obviously an outrider-"Latin" by virtue of its French colonial past and Francophone culture. Parts of the United States might also qualify: Puerto Rico, most obviously; the American southwest, which was once Mexican and which, in the last 60 years, has been re-Mexicanized; and perhaps Miami.
4. Of course, literacy is a complicated question and bald figures often fail to distinguish between reading and writing, varying levels of competence, and the use to which literacy is put.
5. This shift occurred during the Porfiriato (1876–1911), when Monterrey heavy industry grew, while Bajío manufacturing (e. g., the leather industry) stagnated. More recently, the manufacturing output of the Federal District has fallen, while that of the northern border states has grown, thanks to the maquiladoras.
6. I take this emic question to subsume a related but even more tricky problem: when did Latin Americans first *want* to be modern, i.e., acquire aspirations to modernity.
7. Ethnic labels can be attributed on the basis of self-definition (the criterion I stress here); of the opinion of others (neighbors, authorities); or of some supposedly objective criteria (like language, dress, customs). I discuss this problem of attribution—chiefly regarding "Indian" identity—in Knight (1990: 74–5). In asking the question here I have switched "Indian" for "mestizo," since the question "when was Latin American Indian?" could elicit a scornfully simple answer: before ca. 1500.
8. With the caveats expressed in n. 4 earlier. Interestingly, Inkeles and Smith (1974: 252), report that emic answers (i.e., individuals' claims to literacy) are substantially confirmed by follow-up tests.

9. Thomson, this volume. Escalante Gonzalbo (1992) discussing Mexico's nineteenth-century political culture, refers (17, 63, 98, 99) to "the modern spirit," the "threat" (to peasants) of "modernity," "efforts to modernize the Spanish state" and the "construction of a modern state"; however, "modern" and its cognates appear not to have been used by the historical actors themselves; instead, we get "civilization," 14, 268, "barbarism," 57, 92, republicanism, 201, and so on.

10. I add "perhaps" because the second quote refers, rather elliptically, to "Sarmiento's project" coming "into line with emerging visions of modernity."

11. Bernard (1973) confirms this. The only Hispanic example I could think of was the Spanish Falange Español Tradicionalista y de las Juntas de Ofensiva Nacional Sindicalista. The "traditional" bit of this jawbreaker derives from Carlism (and sits rather uneasily alongside the more "modernist" slant of the second half of the label). Latin America, of course, has never nurtured an enduring popular monarchical movement like Carlism.

12. Certainly the term was not used by or about Cárdenas in 1930s Mexico; and, on the basis of lesser expertise, I do not think that either Peronismo or Varguismo were described in these terms by contemporaries.

13. This is sometimes the problem with Cliometrics–economic history informed by econometrics. Its practitioners may ask pertinent questions and deploy sophisticated techniques, but when the historical data—proxies included—are inadequate to the task, the conclusions are questionable. Two consequences follow: Cliometricians either limit themselves to those areas (often quite small) where the data are adequate; or, less creditably, they make a fetish of their theoretical and methodological sophistication—which may not yield much by way of historical explanation but at least impresses onlooking economists.

14. Even this conventional assumption has its critics: Levine (2001) discerns the "roots of the modern world" and "the first phase of early modernization" in the eleventh century, which witnessed a "breaking with antiquity (1, 2, 5); subsequent phases involved the sixteenth-century expansion of European states and capitalism and the nineteenth/twentieth-century "stage of industrialization and mass modernization." At the other extreme, Gran (1996: 337) argues that "modernity as embodied in the capitalist nation-state system began . . . in the 1860s–1880s." There can be few historical/social-scientific concepts which, like modernity/modernization, are capable of generating such vast discrepancies.

15. I mean, of course, that the creation of *Latin* (*sic*) America happened to coincide with the onset of modernity, according to this schema. What existed prior to ca. 1500 was not "Latin," neither was it "modern."

16. Larraín (2000: 14), citing Enrique Dussel. Dussel is not alone in seeing the conquest of the Americas—coinciding, presumably, with the

Reformation—as being bound up with the onset of modernity. I offer an alternative view in what follows.

17. Larraín (2000: 48–69), agrees. However, he does so by virtue of excluding "semi-feudal" Spain (48) from modernizing Europe—the Europe that, supposedly, participated in "the beginning of European modernity" ca. 1500 (4). Yet ca. 1500 Europe—not just Spain—was still dominated by monarchical confessional states allied to a powerful clerisy and Larraín's diagnostic features of modernity (including "reason, progress, political democracy [and] science": 67) were notably absent.

18. This is my own cobbled-together checklist. One of the best, brief (and critical) resumés of modernization theory, its origins and import, is provided by Wolf (1982: 1–13).

19. Inkeles and Smith (1974: 15). For a different set of "institutional orders of modernity," with no less than six helpful diagrams, see Giddens (1987: 141–6, 310–24). Larraín (2000: 15) cites a somewhat different Giddensian menu of modernity. Renegger (1995: ch. 1) offers further definitions and examples.

20. Inkeles and Smith (1974: 19–25). Compare Kahl (1974: 18–21).

21. We could go further and treat "modernity" or "modernization" as "radial" concepts, which do not require common core attributes, yet remain (somehow) recognizable: Collier and Mahon (1993).

22. Israel (2001: 3–4). I make an exception of rationality defined in terms of ends and means, since, in the form of "bounded rationality," it is pervasive and hardly characteristic of Enlightenment modernity: medieval flagellants, e.g., were "rational" in pursuing spiritual well-being and a place in heaven according to the tenets of their faith (i.e., in accord with revealed religion); but they were not rational in the more specific modern, Enlightenment sense. See Freund (1968: 140–3).

23. Israel (2001: 11, 14); Renegger (1995: 45–52); McIntyre (1985), though critical of the "Enlightenment Project," sees "the transition into modernity" as "a transition both in theory and in practice and a single transition at that" (61). Contemporaries, like Voltaire, believed that they were living through an intellectual revolution, Jones (2002: 187).

24. "Practical men who believe themselves to be exempt from any intellectual influence are usually the slaves of some defunct economist. Madmen in authority, who hear voices in the air, are distilling their frenzy from some academic scribbler of a few years back" (Keynes 1936: 383).

25. Israel (2001) does not labor the term ("modernity" does not appear in the book index), but it is central to his magisterial analysis (e.g., 24, 45, 124) and, of course, it is flagged in the book's subtitle: "the making of modernity."

26. In usefully reviewing contrasting definitions, Renegger (1995: 41–2) makes a clear distinction between modernity as a "mood"

(e.g., William Connolly's conception) and modernity as a "sociocultural form," tied to new forms of social organization (e.g., Giddens); the analysis presented here points to the first. However, Renegger then asserts that the former view is "philosophical," while the latter is "historical'; in my view, this is a false dichotomy, the "mood" of modernity being precisely historical and even datable.

27. Israel (2001) stresses both the Enlightenment's spatial reach ("from the depths of Spain to Russia and from Scandinavia to Sicily" 7) and its social penetration (it "profoundly involved the common people, even those who were unschooled and illiterate" 5). Though his chief concern is the content and significance of the *ideas* of the Enlightenment, Israel recognizes the role of "newly invented channels of communication, ranging from newspapers, magazines, and the salon to the coffee-shop and a whole array of fresh cultural devices of which the erudite journals . . . and the 'universal' library were particularly crucial" (vi). For an interesting and highly original Latin American parallel, see Forment (2003), which charts the spread of civic associations and republican citizenship in the nineteenth century.

28. An old but still useful guide is provided by Whitaker (1961).

29. It seems to me (not an expert) that the vulnerable flank of Israel's *Radical Enlightenment* is its social-historical rather than its intellectual-historical claims (see note 27 earlier). It is, of course, notoriously difficult to calibrate the impact of ideas, especially over an entire Continent; but I am more persuaded of the *originality* of Enlightenment modernity—as displayed in the work of a wide range of thinkers and writers—than of their pervasive spatial and social impact.

30. I later draw a passing parallel with Christianity and Marxism: big ideas which, over time, have similarly been selectively used and abused in response to particular interests, incentives, and circumstances.

31. As I noted earlier, "rational" and "rationalized" can mean different things and, by many criteria, Catholicism is neither more rational, nor less irrational than Pentecostalism. Here, I (tentatively) use the term in the same sense as Geertz (1993: 171–2), where Geertz, following Weber, denotes as "rationalized" those religions that are "more abstract, more logically coherent, and more generally phrased"; such religions, I would add, being run by more organized, "professional" and hierarchical clerisies.

32. Cf. Inkeles and Smith (1974: 229–30), who, it seems to me, seek to defend their conflation of "industrial" and "bureaucratic" with "modern" by means of assertive and circular arguments.

33. Shermer (1997); Vyse (1997: 17–8, 213–4), which reveals that in 1994 79 percent of Americans believed in miracles, 72 percent in angels, and 65 percent in the devil (why, especially given the world we live in, is the devil less credible than angels?); and that, of Vyse's Connecticut College students, 69 percent believe that dreams predict

the future, 66 percent that some people are born lucky, and 43 percent that astral projection (the mind or soul traveling independent of the body) occurs.

34. Cf. Talmon (1985); first pubd. 1952, which prefigures more recent Foucaultian thinking, in seeing Nazism, e.g., as the culmination rather than the betrayal of the (European) Enlightenment. The *reductio ad absurdum* comes with "Hitler the philosophe," a "popularizer . . . of "Enlightenment values" : Hollinger (2001: 18).

35. Writing ca. 1973, Inkeles and Smith (1974: 301), conclude, regarding Argentina and Chile, that "the evidence . . . indicates that modern men are likely to favor fundamental change in political and economic institutions, much as they favor basic changes in interpersonal relations and social customs." Chile experienced a coup that year and Argentina three years later. However, that was not quite the "fundamental change in political and economic institutions" that modern men were meant to favor.

36. Greene (2001: 418) notes that "modern" and "tradition" are in common use, "regardless of how thoroughly discredited this dichotomy is in critical intellectual circles." For examples of convincing discrediting, see Wolf (1982: 12–3, 23) and Tipps (1973).

37. For an example of the last phenomenon, Wiarda (2001). On the culturalist turn, see Inglehart (1988: 1203–30).

38. Hollinger (2001) gives some good examples; the list could easily be lengthened.

39. Cf. Charles Baudelaire, "The Painter of Modern Life," in Fascina and Harrison (1982: 23–7); Larraín (2000: 16). The danger of off-the-cuff quotation (especially of poets?) is evident here: Larraín traces the "origins" of modernity to Baudelaire's 1863 essay, where modernity is defined as "the ephemeral, the fugitive, the contingent, the half of art whose other half is eternal and immutable." However, Baudelaire goes on: "Every old master has had his own modernity; the great majority of fine portraits that have come down to us from former generations are clothed in the costume of their own period." Here, like his contemporary Sarmiento, Baudelaire seems to equate "modernity" with "contemporary": modernity is a recurrent feature, not a one-off recent innovation; and it scarcely affords the basis for an elaborate social concept.

40. For examples (Cousiño, Morandé, Veliz, and others), see the critique of Larraín (2000: 66–8, 149–50, 176–9) The use of "baroque" is particularly bizarre, as Larraín rightly points out, when it is coupled with "modernity" in opposition to "enlightened modernity": ibid.: 66–7.

41. From Carol Reed's 1949 film, *The Third Man*.

42. "Influential" but not determinant: consistent with my earlier critique of Keynesian idealism, and tending to a neorealist and political-economy view of U.S. policy, I would locate the main drivers of that

policy in material interests (e.g., oil and corporate profit) and related geopolitical concerns, rather than ideas and ideologies (which "constructivists" would stress). Of course, ideas and ideologies count, and may serve as useful legitimating devices, especially when selling policy to a gullible public.

43. For example, Van Young (2001) stresses the traditional, localist, and communitarian motives of popular insurgents (exemplified in both messianic and "naive monarchist" movements), while Guardino (1996) discerns a political *prise de conscience*, whereby popular rebels espoused new notions of republican representation and citizenship. There is broad agreement among historians, however, that the Catholic Church was seriously divided, with a good many parish priests supporting the insurgency (against what they saw as a secularizing and "Frenchified" [*afrancesado*] regime), while the hierarchy tended to remain loyal to the Crown and the "colonial pact."

44. The final words of Wittgenstein's *Tractatus logico-philosophicus*.

References

Annino, Antonio (1995). *Historia de las elecciones en Iberoamérica, siglo XIX*. Buenos Aires, FCE.

Bernard, Jean-Pierre et al. (1973). *Guide to the Political Parties of South America*. Harmondsworth: Penguin Books.

Clews Parsons, Elsie (1936). *Mitla. Town of the Souls*. Chicago: University of Chicago Press.

Collier, D. and Mahon, J. E. (1993). "Conceptual 'Stretching' Revisited: Adapting Categories in Comparative Analysis." *American Political Science Review* 87: 845–55.

Cosío Villegas, Daniel (1955–65) coord. *La historia moderna de México*, 9 vols. Mexico: Hermes.

Coutinho, Marilia (2003). "Tropical Medicine in Brazil: The Case of Chagas' Disease," in Diego Armus (ed.), *Disease in the History of Modern Latin America*. Durham: Duke University Press, pp. 76–100.

Escalante Gonzalbo, Fernando (1992). *Ciudadanos imaginarios*. Mexico: El Colegio de México.

Fascina, Francis and Charles Harrison (eds.) (1982). *Modern Art and Modernism: A Critical Anthology*. London: Harper & Row.

Forment, Carlos (2003). *Democracy in Latin America, 1760–1900, vol. 1. Civic Selfhood and Public Life in Mexico and Peru*. Chicago: University of Chicago Press.

Freund, Julien (1968). *The Sociology of Max Weber*. Harmondsworth: Penguin Books.

Geertz, Clifford (1993). *The Interpretation of Cultures*. London: Fontana.

Giddens, Anthony (1987). *The Nation-State and Violence*. Berkeley: University of California Press.

Gran, Peter (1996). *Beyond Eurocentrism. A New View of Modern World History*. Syracuse: Syracuse University Press.

Greene, Alison (2001). "Cablevision(nation) in Rural Yucatán: Performing Modernity and Mexicanidad in the Early 1990s," in Gilbert M. Joseph, Anne Rubinstein, and Eric Zolov (eds.), *Fragments of a Golden Age. The Politics of Culture in Mexico Since 1940*. Durham: Duke University Press.

Guardino, Peter (1996). *Peasants, Politics and the Formation of Mexico's National State: Guerrero, 1800–57*. Stanford: Stanford University Press.

Guardino, Peter (2005). *Popular Political Culture in Oaxaca, 1750–1850*. Durham: Duke University Press.

Guerra, François-Xavier, Annick Lempérière, et al. (1998). *Los espacios políticos en Iberoamérica*. Mexico: FCE.

Harris, Marvin (1976). "The History and Significance of the Emic/Etic Distinction." *Annual Review of Anthropology* 5: 329–50.

Harrison, Laurence E. and Huntington, Samuel P. (eds.) (2000). *Culture Matters. How Values Shape Human Progress*. New York: Basic Books.

Hollinger, David A. (2001). "Enlightenment and the Genealogy of Cultural Conflict in the United States," in Keith Michael Baker and Peter Hanns Reill (eds.), *What's Left of Enlightenment?* Stanford: Stanford University Press, p. 18.

Huntington, Samuel P. (1996). *The Clash of Civilizations and the Remaking of World Order*. New York: Simon and Schuster.

Inglehart, Ronald (1988). "The Renaissance of Political Culture." *American Political Science Review* 82: 1203–30.

Inkeles, Alex and David H. Smith (1974). *Becoming Modern. Individual Change in Six Developing Countries*. Cambridge: Harvard University Press.

Israel, Jonathan I. (2001). *Radical Enlightenment. Philosophy and the Making of Modernity, 1650–1750*. Oxford: Oxford University Press.

Jones, Colin (2002). *The Great Nation. France from Louis XV to Napoleon*. London: Penguin.

Kahl, Joseph A. (1974). *The Measurement of Modernism. A Study of Values in Brazil and Mexico*. Austin: University of Texas Press.

Keynes, J. M. (1936). *The General Theory of Employment, Interest and Money*. New York: Harcourt Brace.

Knight, Alan (1990). "Race, Racism and Indigenismo: Mexico 1910–14," in Richard Graham (ed.), *The Idea of Race in Latin America, 1870–1940*. Austin: University of Texas Press.

——— (1994). "Popular Culture and the Revolutionary State in Mexico, 1910–40." *Hispanic American Historical Review* 73: 393–444.

——— (1998). "Populism and Neo-Populism in Latin America, Especially Mexico." *Journal of Latin American Studies* 30(2): 223–48.

——— (2002). *Mexico: From the Beginning to the Spanish Conquest*. Cambridge: Cambridge University Press.

Larraín, Jorge (2000). *Identity and Modernity in Latin America*. Cambridge: Polity Press.

Levine, David (2001). *At the Dawn of Modernity. Biology, Culture and Material Life in Europe after the Year 1000.* Berkeley: University of California Press.

McIntyre, Alasdair (1985). *After Virtue. A Study in Moral Theory,* 2nd edn. London: Duckworth.

MacLachlan, Colin M. and Jaime E. Rodríguez O. (1980). *The Forging of the Cosmic Race, A Reinterpretation of Colonial Mexico.* Berkeley: University of California Press.

Melville, Eleanor G. K. (1994). *A Plague of Sheep: Environmental Consequences of the Conquest of Mexico.* Cambridge: Cambridge University Press.

Nicholls, Peter (1995). *Modernisms: A Literary Guide.* London: Macmillan.

Redfield, Robert (1930). *Tepoztlán, a Mexican Village.* Chicago: Chicago University Press.

Renegger, N. J. (1995). *Political Theory, Modernity and Postmodernity.* Oxford: Blackwell.

Sarmiento, Domingo F. (1961). *Facundo. Civilización y barbarie.* New York: Doubleday.

Scott, James C. (1998). *Seeing Like a State. How Certain Schemes to Improve the Human Condition Have Failed.* New Haven: Yale University Press.

Shermer, Michael (1997). *Why People Believe Weird Things.* New York: Henry Holt.

Talmon, J. L. (1985). *The Origins of Totalitarian Democracy.* Boulder: Westview Press, first pubd. 1952.

Tipps, Dean C. (1973). "Modernization Theory and the Study of National Societies: A Critical Perspective." *Comparative Studies in Society and History* 15(2): 199–226.

Van Young, Eric (2001). *The Other Rebellion: Popular Violence, Ideology and the Mexican Struggle for Independence, 1820–21.* Stanford: Stanford University Press.

Viñas, David (1994). "Sarmiento: Madness or Accumulation," in Tulio Halperín Donghi, Ivan Jaksic, Gwen Kirkpatrick, and Francine Masiello (eds.), *Sarmiento. Author of a Nation.* Berkeley: University of California Press.

Vyse, Stuart A. (1997). *Believing in Magic. The Psychology of Superstition.* Oxford: Oxford University Press.

Whitaker, Arthur (1961). *Latin America and the Enlightenment.* Ithaca: Cornell University Press.

Wiarda, Howard J. (2001). *The Soul of Latin America. The Cultural and Political Tradition.* New Haven: Yale University Press.

Wilkie, James W. (1970). *The Mexican Revolution: Federal Expenditure and Social Change since 1910,* 2nd edn. Berkeley: University of California Press.

Wolf, Eric R. (1982). *Europe and the People without History.* Berkeley: University of California Press.

Part II

Views from Literary and Cultural Studies

Chapter 5

When Was Peru Modern? On Declarations of Modernity in Peru

William Rowe

Para abrir por fin rendijasen la pared del tiempo.
E. A. Westphalen

Knowledge, like history, is incomplete.
Georges Bataille

In this essay I have made a number of "cuts" in the temporal continuum of what is thought of as the modern period in Peru, and in the ordering and presentation of materials I have sought to embody the nonlinearity that my argument proposes as necessary for understanding what is problematic about the idea of modernity in Peru. To think of them as cuts in the continuum of history is misleading, since this would simply be the continuum that history is imagined to consist of. In fact they are not cuts in a preexisting continuum, given that each one constitutes of itself a particular temporality: that is, each one displays the work of constituting a temporality. The type of effect they produce can be summed up in the proposition that there is no such thing as a single temporal continuity, not even a single continuity with several strands, but rather, various cuts or moments in which the time of the modern is constituted in Peru. By constituted I mean that its components and the relationships between them become recognizable or readable. That several such moments can be identified—and my series could no doubt be added to—points up the partial and contested nature of the modern in Peru.

The essay is divided into sections. Each one explores a scene in which temporality is generated. The intention is that they can be read in any order, since they do not make up a linear series, and are not accumulative. The idea is that they should work like, that is, be readable as mobile sections that intersect with each other. At the same time each of these scenes makes the relation between the past and the present legible in particular ways. Here I have in mind Walter Benjamin's statement, in the Arcades Project:

> What distinguishes images from the "essences" of phenomenology is their historical index [. . .] The historical index of the images not only says they belong to a particular time; it says, above all, that they attain to legibility only at a particular time. And, indeed, this acceding "to legibility" constitutes a specific critical point in the movement at their interior. Every present day is determined by the images that are synchronic with it: each "now" is the now of a particular recognizability. (Benjamin 2002: 462–3)

Manuel González Prada and the Declaration of Modernity

Modernity in Peru has usually been asserted by declaration of one type or another. There was González Prada's famous declaration, after the defeat in the War of the Pacific, of "los viejos a la tumba, los jóvenes a la obra" (González Prada 1966: 64) or Mariátegui's slogan, "hacer país," or Leguía's hailing the twentieth century with public works in stone and stucco that incorporated indigenous design motifs, or the later declarations in concrete during the regimes of Odría and Belaúnde that hailed the expansion of capitalist modes of production in the post second world war period. All of these moments included a clearing action—both discursive and spatial—and all were contested by counterstatements and counter-scenarios. González Prada's speech in the Politeama theater (1888) cleared the language and its scenes of enunciation of all that might bespeak a colonial inheritance; it outlawed, in the name of science, the obfuscations of scholastic thought, and in its passionate announcement of moral crisis, silenced previous conversations complicit in defeat and failure and disqualified their interlocutors. The past, he said, consisted in "teología y metafísica," and the future belonged to "la Ciencia positiva" (González Prada 1966: 63).

One way to uncover these suppressed conversations is to read Ricardo Palma's *Tradiciones peruanas*, and hear the voices in the streets of Lima, which still carried the moral ambiguities and subterfuges of colonial society, the superficial obedience to hierarchical rules of discourse that in real life were simultaneously obeyed and eluded. Palma shows the popular black saint of sixteen-century Lima,

Fray Martín de Porres, eluding the double bind of colonial authority by turning it against itself. As someone who, in Palma's words, "hacía milagros con la facilidad con que otros hacen versos," Fray Martín was forbidden by his superior to do any more miracles ("tuvo que pro-hibirle que siguiera milagreando (dispénsenme el verbo)"). Palma then cites one of his biographers, in order to have him say to a worker who fell from scaffolding 8–10 meters from the ground, " Espere un rato, hermanito!" And, as Palma narrates it, "el albañil se mantuvo en el aire haste que regresó fray Martín con la superior licencia" (Palma 2000: 566). This extraordinary faith in the authority of the word is deployed in order to make the word of authority laughable. Thus González Prada's ground-clearing rhetorical sweep was contested by that sense in Palma's narratives of a colonial past continuing to speak—indeed, by the pleasure taken by readers in their skill in navi-gating the ironies of his narratives. In fact it could be argued, against González Prada, and against subsequent uses of his tone (e.g., by Sebastián Salazar Bondy in *Lima la horrible* (1964)) that the ability to turn colonial rules of discourse against themselves is actually an early sign of Peruvian modernity.

The clearing action of these declarations of modernity that I have mentioned was, as I have indicated, spatial as well as rhetorical. The creation of wide avenues and of residential suburbs under Leguía involved—as did Haussmann's Paris of the nineteenth century or Moses's New York in the twentieth century—the clearance of slums. The ways in which other visual experiences contested this moderniza-tion of the urban landscape of Lima were not as far as I know docu-mented at the time, but they can be guessed at through a reading of Julio Ramón Ribeyro's early stories, such as "Los gallinazos sin plumas" (1955), which from the mid-1950s to the early 1960s dis-played ongoing scenarios of colonial abjection existing alongside the spirit of free enterprise. The final clearing away of colonial Lima as slum coincided, perhaps not unexpectedly, with the restoration of fine colonial buildings such as the Casona of the University of San Marcos in the Parque Universitario or the clearing and renovation of the Plaza de Armas area. In other words, the colonial can become again more visible once its social scars become less so. With the González Prada cycle complete, nostalgia becomes heritage.

What about Mariátegui's calls for the creation of a genuine nation? The most obvious counterarguments came perhaps from José de la Riva Agüero, in the form of the assertion that there already was a nation, consecrated by the blood of its martyrs. It was of course a protest against the consecration of Peru to the Sacred Heart of Jesus that furnished the occasion for the demonstration led in 1923 by Víctor Raúl Haya de la Torre which most historians take to be the marker that initiates that key narrative of twentieth-century Peruvian history, the entry of the masses into politics. The situation was in fact a great deal more complex than these quick indications of the symbolization of epochs are able to convey. Haya de la Torre, for example, while sharing Mariátegui's concern with the creation of a modern nation, did not go along with his belief that Peru would not need to pass one by one through the stages of European history (from feudalism to capitalism to socialism) because the character of Peruvian temporalities and the relationship between them was different. As Alberto Flores Galindo stated, Mariátegui's vision of Peruvian history was marked by "su recusación del progreso y su rechazo de la imagen lineal y eurocentrista de la historia universal" (Flores Galindo 1994: 434). *Recusar* claims an act of justified refusal. Mariátegui's thinking engenders the possibility that Peruvian history would no longer have to be conceived as a lack, that is, as inhabited by uchrony (Chocano 1987: 43–60) by what would have happened if . . . Peru had had a proper bourgeoisie or any other of those ingredients of progress (or, indeed, of development?). However, in order to sustain what Flores Galindo calls "la ruptura con esa imagen de una historia universal impuesta por los europeos a todos los países atrasados," there were a number of implications. One of them was not just a break with academic historiography, but also with the dominant version of Marxism as expressed by Engels and later Stalin, which upheld the "esquema clásico que partiendo del comunismo primitivo, seguía por el esclavismo, desembocaba en el feudalismo y llegaba al capitalismo."

The other condition for overcoming the definition of Peruvian history as a lack, as what *might* have happened, was the existence of a political movement capable of breaking with the temporality of progress and instituting a different one. Jorge Basadre, the most important historian of the Republic, imagined Peru as a unity after crossing the Andes in an aeroplane. However, the transformation of Peruvian space by modern technology, which had also occurred previously with the nineteenth-century building of railways and, even more so, with Leguía's program of road building in the second and third decades of the twentieth century, was not of itself sufficient to wrest the temporal imagination away from the fascination and power of linear programs of modernization, which placed Peru in a situation of deficit. Basadre recounts his flight over the Andes in his 1931 book, *Perú: problema y posibilidad*, where he also wrote: "en el Perú no había sino vida local. Precisamente no existía la vida nacional. La solución está [. . .] en forjar por medio del localismo, la autoconciencia de la nación, que no existe" (Basadre 1992: 139). The sequence of verbal tenses is curious: "no había, no existía," followed by "está" and "no existe." There is the past continuous, without definite end or beginning ("no había," "no existía") and then an awkward switch to the present tense: awkward because normal syntax requires some completion of the (imperfect) past in order for there to be a switch to the present. The normal progression would be from imperfect to preterite ("no había . . . la solución fue . . .). But the past goes on, without completion, and the narrative switches to the future whose content is that which does not exist in the present. In other words, this future of what has to be done[1] does not complete the past. The past continues, in the provinces, without definite end or beginning, exactly as it does in Valdelomar's iconic provincial story, "El caballero carmelo" (1918).

The temporal gap (or aporia) is also a historiographical one. The grammar of time operating in Basadre's sentences is symptomatic of their theme: the lack of geographical articulation translates into temporal disconnection. The theme is the provinces versus the capital: provinces from which the intellectuals who in the 1920s reimagined Peru had come: "la provincia vivió sólo para votar para un remoto y abstracto parlamento, para recibir autoridades políticas y para seguir una vida sórdida. Todo el resto de la vida del país fue, según mandaron las Constituciones, vida nacional, vida de la capital." In this passage there is a single narrative time, told from the homogenizing point of view of Lima, within whose temporality the provinces are included: the preterite embraces both Lima and the provinces ("la provincia vivió . . . todo el resto de la vida del país fue . . ."). Here there is no gap in the discursive logic of time—except of course in the one that, ironically, the logic of constitutions papers over, that is, the fact that the provinces, as we know, were leading a completely different life—"we" here being Basadre's implied reader, a migrant from the provinces like himself. And that suggests another way of accounting for the awkwardness of the passage previously quoted: it's as though Basadre were narrating somebody else's discourse as indirect speech and then switching to direct speech: "no había eso, no existía lo otro pero—dijo—la solución está en lo les voy a decir ahora." The trouble is, though, that this somebody else would be himself. What I am suggesting, in other words, is that here we have Basadre the provincial migrant hearing himself speak as the historian now installed in Lima, in the place of constitutions, parliaments, and other instances of discursive authority. Basadre's writing owns up to the scene of the historian's authority and as it does so it indicates a certain awkwardness and unease.

Basadre had registered the difficulties of producing historiographical narrative in Peru in his first chapter: "la síntesis social peruana—hay que repetirlo—no se ha realizado aún. El pasado peruano no es algo colmado ni admirable; y el Perú sigue siendo una serie de compartimientos estancos, de estratos superpuestos o coincidentes, con solución de continuidad" (Basadre 1992: 12). The paragraph from which I quote is headed "El porvenirismo en la historia peruana." Where, then, to find the continuity that historiography needs in order to produce itself? When he comes to the final pages of the book, in a section titled, "Balance final," he asserts that the only positive continuity is provided by the unrealized thing called Peru: "Y el Perú, con todos estos males y sus amenazas coincidentes, ha sobrevivido como si su mensaje aún estuviera por decir, como si su destino aún no estuviese liquidado, como si llevase consigo una inmensa predestinación" (Basadre 1992: 152). At which point, that is, in the light of the continuity provided by that imagined future, he adds, in a sentence given dramatic emphasis by the fact that it forms a paragraph on its own: "No ha habido integración en los estratos sociales pero sí una marcha hacia esa integración." Thus, finally, in the light of this transcending predestination, it can be written that there has been significant movement forward, there has been progress.

Beside his vision of a future of "justicia social," to be achieved through socialism, Basadre places, again in the final pages of the book, what he calls his "razones para dudar": "Taras, culpas y errores hacen incrementar los factores de disociación [. . .] carecemos de victorias y de grandes hombres. Las estatuas de los mejores podrían empezar con torsos robustos esculpidos por finos cinceles, concluidos luego rudamente, a machetazos."[2] To this rhetoric of uchrony, he adds, in Gonzalezpradesque tone, that to resolve the problems of Peru "en beneficio de las masas que constituyen el auténtico país," will be the task of "las nuevas generaciones" (Basadre 1992: 156). The language of what might have happened is thus projected into the future.

From the point of view of its implied scene of enunciation and, crucially, the split within that scene between provincial temporality and that of the capital city, Basadre's *Perú: problema y posibilidad* can be compared with a number of other scenes that display the contexts of discourses of modernity. These contexts constitute and are constituted by particular language games—to use Wittgenstein's term—or, in Roland Barthes's terms, are ruled by different referential codes. The rest of this chapter will set out a number of these scenes.

The order in which they are given is only one of the possible orders in which they can be read. The intention is that they can be read like mobile sections, rather than as stages in a sequence.

Eguren's Motor Car

This is perhaps the most incongruous of the various scenes. José María Eguren (1874–1942) was born in an hacienda in the outskirts of Lima; after the War of the Pacific (1879–83) the family moved to Barranco, at that time a small coastal town just outside Lima. In a poem called "La comparsa," published in 1911, he wrote:

Allí van sobre el hielo las figurantas
sepultando en la bruma su paranieve,
y el automóvil rueda con finas llantas,
y los ojos se exponen al viento aleve.

A *comparsa* is a procession of masked figures, and *figurantas* includes reference to the theatrical effect of such figures. "Paranieve" is explained in a popular edition of Eguren's poetry as "sombrilla para evitar la nieve (como parasol)" (Eguren 1961: 65). What are ice and snow—not to speak of an umbrella for snow—doing in the writing of a Lima poet? It's the modern taste for the gothic: Poe, possibly via Baudelaire. Martín Adán, another Lima poet, is vehement about its inappropriateness: "A estar a mi noticia, en ninguna poesía en español se nota más en extensión y en intensión la falta real de septentrión feliz, donde lo gótico cumplió su curso natural" (De la Fuente Benavides 1968: 352). His point, none the less direct for the erudite word *septentrión*, is that gothic scenarios only work properly in the north—which doesn't get us any nearer to an answer, except to say that the ice and snow aren't really there.

What about the motor car with its "finas llantas"?—not the usual word for a car tire (= neumático in Spain)—which makes *llantas* an example of the different and usually more rapid creation of modern technical terms in Latin America: the faster absorption, in the language, of technological modernity.[3] The car, though slowed down to the speed of strolling masked figures, is there without adjectives, a clear sign of modernity. But it is not actually driving on the snow: in fact the ground does not appear. This covering-over of the ground can be read as one of the conditions by which Eguren also achieves a clearing action, a visual declaration of epoch. The gothic is a mask, a semi-diaphanous veil—like the mist (*bruma*), the white mist of Lima, which is there for some six months of the year and which, unlike northern mist, incorporates a tropical glare, making for a diffuse brightness, without clear outlines. Eguren uses this effect for his system of the visual—he was a photographer and watercolorist—which works through veilings and unveilings. The motor car, then, is subject to the subtle variation of visibilities, to fluctuating light. The snow in this context is a type of veil, as in the lines "y al poniente fluctúa luz incolora, / y los méganos ciñe la nieve obscura," where *méganos* is an unusual word for sand dunes, which are, according to Eguren's language game, the sand dunes that surround Lima to the south, east, and north, even more in the time of Eguren, when they were not the site of "pueblos jóvenes."

The visual in Eguren's poems is often completely static. One of his favorite figures is the classical statue in a niche. The slow walking procession—and not mechanized transport like the tram, which ran from Barranco to Lima from 1896—gives the speed of movement in the poem that I have been quoting. The modern enters, but as a semi-static figure, disengaged from the new technologies of speed. There is something still colonial about the atmosphere. As Adán comments, "Lima [. . .] en la infancia de Eguren conservaba mucho de colonial, luego desaparecido [. . .] Era ciudad con semblante propio, [. . .] triste y sombrío," and then adds (he is writing in the 1930s) "como la padecemos todos, sin teoría y sin reparo, deleznable y tenaz, uniforme y varia, modernista y ruinosa, tradicional y novelera."[4]

Martín Adán's Mule

Each scene of enunciation is related to a particular type of visibility just as, conversely, each regime of the visible interacts with what is sayable. Taken as a whole, each scenario of visibility/scene of enunciation can be taken as a type of reading machine, producing different types of readability.

In Martín Adán's (1908–85) novel *La casa de cartón* (1928), the urban scene is cinematic in its mode of visibility:

> Y estos autos, sucios de prisa, de orgullo, de barro . . . Los ficus hacen crecer las casas en sus espejismos de follajes de lodo y musgo, casi agua, casi agua, agua por arriba y abajo [. . .] Gorriones, saltamontes. Uno mismo abre los ojos redondos, ictiologizado. En el agua, dentro del agua, las líneas se quiebran, y la superficie tiene a su merced las imágenes. No, a merced de la fuerza que la mueve. Pero de lo mismo, al fin y al cabo. Pavimento de asfalto, fina y frágil lámina de mica . . . Una

calle angostísima se ancha, para que dos vehículos—una carreta y otra carreta—al emparejar, puedan seguir juntos, el uno al lado del otro. Y todo es así temblante, oscuro, como en pantalla de cinema. (Adán 1971: 22–3)

Here there are several ways in which the visibility of things becomes similar to that which a cinema screen produces. Unlike Eguren's, Adán's vehicles—motor cars and carts—move with some speed, fast enough to come in and out of the frame. In fact, the eye moves with the vehicles, rather than being static: this is a cinematic eye. At the same time the medium, the medium within which things pass in and out of visibility, is figured as water. We start with a montage of the leaves of fig trees and houses and then of sparrows and grasshoppers. The transition, from the large to the small and very small, is effected through the medium of water: the fish's eye does not see objects in perspective: the geometric grid is broken (as with a fish-eye camera lens?), everything is close up. Together with the breaking of those lines of perspective that placed the old-fashioned, Cartesian spectator outside the frame, the water conveys the haptic effects of the visual medium: just as Bill Viola figures the video medium in his installations by placing screens under water or by showing images of himself floating under water. But Adán is also concerned with the site or screen on which the images occur: the surface of the water is like a cinema screen. What causes the images to move? The first answer given is the surface of the water. But that is then seen as insufficient, since it does not account for the force that runs through them, the depth. Adán then moves to a synthesis: the force and the surface are the same. Thus the dynamics of movement are also in the pavement, itself a surface ("lámina de mica"), layered with other surfaces; the novel as a whole is concerned with surfaces and movement, stasis and duration, suicide and time. The passage ends with a classic filmic effect: the relative movements of two vehicles composed into one, through montage, that is, montage within the shot rather than montage between shots; here the street, the city itself, brings about the montage, just as, in a later passage, "el panorama cambia como una película desde todas las esquinas."[5] The impact of film on pre-cinematic visuality corresponds, in Adán's novel, with the impact of modernization on the city.

The situation of the spectator in the cinema and the conversion of that into a way of reading is completed by the sound track:

Desde un millón de puntos de vista, en un tango largo como un rollo de película, filmaba una victrola a cámara lenta el balneario—amarillo y

desolado como un caserío mejicano en un fotofolletín grotesco de Tom
Mix—Y, detrás de todo, el mar inútil y absurdo como un quiosco en la
mañana que sigue a la tarde de gimkana. (Adán 1971: 35)

The preelectronic gramophone, which of course is wound up like an
early cine camera, films the town. The music also produces the composition of the visual, including its particular speed. The sound of
modern life becomes its sound track. The forms of expression of the
technology are, ironically, foreign (tangos and cowboy pictures), but
the form of the content is Peruvian.

Walter Benjamin, in his "Small History of Photography," observes
that when a new technology first appears it takes on an earlier form.
This was the case, for example, with wrought iron, which at first was
used not to create new, previously impossible, types of architectural
structure but to carry on a previous decorative style. Or, to cite a more
famous example, photography, when it began, imitated certain features of painting.[6] The spinning mule was the name for an early piece
of technology that contributed to the Industrial Revolution in Britain.
Martín Adan's mule is a mule in a street in Barranco:

La tarde proviene de esta mula pasilarga, tordilla, despaciosa. De ella
emana, en radiaciones que invisibiliza la iluminación de las tres pos-
meridiano y revela el lino de la atmósfera—pantalla de cinematógrafo,
pero redonda y sin necesidad de sombra—; de ella emanan todas las
cosas. Al fin de cada haz de rayos—una casa, un árbol, un farol, yo
mismo. (Adán 1971: 86)

On the mule's back is projected the cinematic substance of reality:[7]
the technology resituated on the back of the residue of the premodern
in Peru. The context of the diffuse, mid-day, tropical light of Lima,
unlike the dark and smoky cinema[8] makes the beams of projected light
invisible, becomes itself a visual surface or surface of the visible,
whose color and tactility is figured in "lino" (linen). The passage
continues: "Esta mula nos está creando al imaginarnos. En ella me
siento yo solidario en origen con lo animado y lo inanimado" (Adán
1971: 86-7). The writing continues into a parody of creation narrative: "A cada paso de la mula—paso dúplice, rotundo inalterable de la
eternidad, predeterminado por un genio divino—tiembla mi ser al
destino inconocido." The scenario—which is both appearance and
origin—is technologically modern, yet its frame and composition are
that of the simultaneously modern and non-modern that characterize

Peru of the 1920s. The conditions of visual and discursive meaning are Peruvian even though their technological media were both invented and first configured elsewhere. The fact that Adán's novel displays and reads these disjunctions from Lima is one of the reasons why it would not be accurate to call it a cosmopolitan novel. Adán's mule articulates within the social space of Lima a modern technoscape of information.

The Modern as Conversation

If modernity is characterized as the experience of evanescence, this makes it possible to conceive of a range of alternative modernities, each specific to given geographical sites. This is the strategy of thinking advocated by Dilip Gaonkar in his introduction to *Alter/Native Modernities*, a special issue of the journal *Public Culture*, published in 1999. Part of his concern is to criticize "acultural" theories of modernity, that is, those operational theories of modernity as transformations that "any traditional culture could undergo."[9] It is important to note in those theories the style of neoliberal—and, currently, neoconservative—thought, with its superstition that there is only one type of reason, because to do so is to place on the table the fact that all modernities stand in some particular relation to global geopolitical forces. Thus for example, González Prada's version of Peruvian modernity would need to be read in relation to the beginnings of U.S. domination of the Pacific and with that, as Luis Rebaza has pointed out, the advocation of U.S. models of modernity. And that brings in another reason for detecting the voice of neoconservatism: if neoconservative thought uses reasons of human rights (e.g., of "the freedom deficit," as Dr. Condoleeza Rice called it in her inauguration speech) to justify military invasion of sovereign territories, then political thought that opposes such types of action needs, as Costas Douzinas has argued, to return to the problem of sovereignty. In other words, the modernity imposed by "revolution from above" also has to be part of the discussion.

Gaonkar's advocacy of alternative modernities relies on the assertion of the modern as a site where "creative adaptation" takes place. This "is not simply a matter of adjusting the form or recoding the practice to soften the impact of modernity; rather, it points to the manifold ways in which a people question the present. It is the site where a people 'make' themselves modern, as opposed to being 'made' by alien and impersonal forces" (Gaonkar 1999: 16). This version of the modern depends upon a particular theorizing of modern temporality as governed by evanescence. Gaonkar draws on the writing of Baudelaire in order to propose that modernity is to be found "at the crossing where the fugitive materiality of the life-world impinges on a sharpened consciousness of the present," adding that "nowhere is that crossing more vivid and dramatic than in the life and work of a modern city, such as Baudelaire's Paris." One of the texts he has in mind is Baudelaire's *The Painter of Modern Life*, from which he quotes the following sentence: "Modernity is the transient, the fleeting, the contingent."

If we turn from visual culture to discursive forms, conversation can be taken as an evanescent form of discourse, in that it is open to present events outside itself. Unlike Gonzalez Prada's speeches, for example, it takes place in such a way as to be open to the occurrence of everyday life. As Baudelaire wrote, "Nearly all our originality comes from the stamp that *time* impresses upon our sensibility." Conversation also composes itself as it goes along. It has, because of this fabrication of present time, become the dominant form of TV news (contrast the older, more declarative forms of cinema newsreels). Jesús Martín-Barbero calls the time of TV news the "autistic present," in the sense that it promotes amnesia (Martín-Barbero 2002: 1).

However, Mario Vargas Llosa's use of conversation in *Conversación en la Catedral* (1969), while in thrall to evanescence, also points up its limitations when used as a theory of modernity. Every conversation in this novel has been preempted by a previous one. As the conversation, which is the novel unfolds, a significant proportion of the words, tones, phrases, paragraphs that make it up turn out to have been uttered previously. The immediate circumstances reveal themselves to be nonimmediate. As Walter Benjamin argued, one can only get a critical understanding of the present through grasping the constellation it makes with the past (Benjamin 2002: 462).

The technical device of placing one conversation inside another inside another had already been initiated in Vargas Llosa's second novel, *La casa verde* (1966). Although the later *Pantaleón y las visitadoras* (1973) is the one that became what was probably the most successful, visually, of the film adaptations of Vargas Llosa's novels, *La casa verde* is the most filmic of his novels. If Conrad imagined his novel, *The Nigger of the Narcissus* (1897) as a fully visual experience, modeled on the painting of Monet, *La casa verde* presents itself as a fully filmic novel in that each scene moves with the pulse of screen images accompanied by a sound track that immediately bursts crackling into life, without that selection of significant detail (both visual and auditory) that Conrad was a master of. Here, for comparison, are the opening paragraphs of chapters from the two novels:

> MR. BAKER, chief mate of the ship *Narcissus*, stepped in one stride out of his lighted cabin into the darkness of the quarter-deck. Above his head, on the break of the poop, the night-watchman rang a double stroke. It was nine o'clock. Mr. Baker, speaking up to the man above him, asked:—"Are all the hands aboard, Knowles?"
> The man limped down the ladder, then said reflectively:-
> "I think so, sir. All our chaps are there, and a lot of new men has come. . . . They must be all there."[10]
> SONÓ UN PORTAZO, la superiora levantó el rostro del escritorio, la Madre Angélica irrumpió como una tromba en el despacho, sus manos lívidas cayeron sobre el espaldar de una silla.
> -¿Qué pasa, Madre Angélica? ¿Por qué viene así?
> -¡Se han escapado, Madre! –balbuceó la Madre Angélica-. No queda ni una sola, Dios mío.
> -¿Qué dice, Madre Angélica—la Superiora se había puesto de pie de un salto y avanzaba hacia la puerta-. ¿Las pupilas? (Vargas Llosa 1966: 23)

Nevertheless, despite the strong force of the visual as narrative occurrence, and the use of cinematic flashbacks (i.e., that are not discursively mediated),[11] the way Vargas Llosa's novel welds together the varied components of its long timespan (1920–60) depends equally if not more on verbal memory. The narrative fragments are held together by the reader's memory of two factors: the underlying, semi-mythical story of the "Casa Verde" itself, and the conversations that cut backward and forward in chronological time. If read as a response to Basadre's sense of the failure of Peru to come together into a unified history, *La casa verde*, which embraces the national territory from the jungle to the coast, over a timespan of 40 years, succeeds in bringing Basadre's "compartimentos estancos" into significant relation with each other. But it does so without imposing a single master-narrative, such as that of progress or of the entry of the masses into history. It places the semi-mythical past of the coastal provinces alongside the violence of commercial capital and the logistic reach of army and police as organs of state and territorial but not social coherence.

With *Conversación*, on the other hand, rather than combining filmic cutting and more traditional verbal narrative, Vargas Llosa fully transfers the capabilities of film montage to the material of conversation itself. The novel consists of a single conversation that passes via unmediated cuts through continual modulation by time and place.

There are scenarios inside scenarios, interlocutors inside interlocutors, leading both "backward" and "forward" in time. And what is placed under continual modulation is not just the theme but the language games that constitute each of its moments. In other words, the meaning of place is not so much the geographical dispersal of Peru, as in *La casa verde*, but the changing meanings of power relations as the conversation moves across the huge differentials of social power. It offers one of the most totalizing scenes of Peruvian modernity. The conversation takes place in the 1940s, during the time of Odría's dictatorship, yet the effect produced by the discovery that every conversation contains a previous one offers a corrective to the emptying out of time by unmediated presentness. Its greatest contradiction is in the fact that the main relationship with the past is one of pathos, through an evocation of lost innocence. But if pathos is a weak relationship with the past, little capable of grasping the discontinuous time of modernity, the formally overdetermined device of discontinuous conversation serves to counter the univocal lure of pathos. If pathos exhausts and closes the past, the continually cut and modulated conversation has an opposite effect of openness. If Valdelomar's classic early twentieth-century story, "El caballero Carmelo" (1918) offers a nostalgic past severed from the present, that is one side of the experience of Peruvian modernity. *Conversación* exposes what is repressed by that image: the way in which the present of modernity continually breaks up and relocates that premodern past.

Before the Walls of Cusco

The scene before the Inca walls of Cusco, where the young protagonist of Arguedas's *Los ríos profundos* (1958) contemplates the irregular stonework, is one of the "foundational" scenarios of modern Peruvian literature. It is a scenario in which the native past, its cultural and political order, becomes readable. But the question is how are they rendered readable, and to what extent? Clearly these are questions that Peruvian historiography has not been able to escape from.

Arguedas's Cusco is a rereading of el Inca Garcilaso's.[12] El Inca's account of the society of the Incas, carried by Simón Bolívar in his saddlebag during the campaign of emancipation from Spanish colonialism, was itself perhaps the first document of Peruvian modernity. Aníbal Quijano argues that although "la modernidad como categoría se acuña [. . .] en Europa y particularmente desde el siglo XVIII [. . .] el proceso de producción de la modernidad tiene una relación directa y entrañable con la constitución histórica de América Latina"

(Quijano 1988: 10–1). His assertion is that the discovery of America, insofar as it produced "una profunda revolución en el imaginario europeo," lies at the basis of the imagination of modernity as a temporal category:

> se produce el desplazamiento del pasado, como sede de una para siempre perdida edad dorada, por el futuro como la edad dorada por conquistar o por construir.
> ¿Cómo se podría imaginar, sin América, el advenimiento de la peculiar utopía europea de los siglos XVI y XVII en la cual ya podemos reconocer los primeros signos de una nueva racionalidad [. . .]
> Y el surgimiento de esas específicas utopías puede ser reconocido como el primer momento del proceso de constitución de la modernidad. Sin el nuevo lugar del futuro en el imaginario de la humanidad, la mera idea de modernidad sería simplemente impensable. (Quijano 1988: 12)

In this context, el Inca's *Comentarios reales* were the first decisive attempt to make Inca society readable to the utopian imagination. And a major part of what makes his book so decisive is that it goes further than Thomas More's *Utopia*, by making available not just the social rationality of an American society but also its universe of signs: its grammar, its lexicon, its myths, and theology, its transformation of the land into networks of symbols.

A prime example of the latter is the chapter on *huacas*, about which el Inca is at great pains to say that they were not gods: the early chroniclers, who did not know the language, made the mistake of calling *huacas* gods. Thus el Inca's insistence on reading the signs, or rather, on making them readable. Rocks and stones are one of the main forms taken by *huacas*: both in that they can travel across the land (i.e., in mythology), in their natural shape, and in the shaping of them and the inscriptions made on their surface. Thus Arguedas's Ernesto on what the Spanish had done to the stones that form the Inca wall he is contemplating: "golpeándolas con cinceles les quitarían el 'encanto' " (Arguedas 1983: 17). Arguedas, in a sense, takes over from where el Inca's history of Peru had left off. But only in a sense, because the scene before the walls, in one of its possible readings, casts radical doubt upon whether the original "encanto" can actually be experienced in twentieth-century Peru.

The temporal implications of Arguedas's scene are complex and conflictive. If, as Quijano suggests, the dominant form of modernity became, in the nineteenth and twentieth centuries, that of instrumental reason, the sheer irregularity of the Inca stonework, its resistance to Renaissance-gridded space, means that if native culture is to be brought into and against occidental modernity, then an alternative rationality, which is not that of instrumental reason, is being invoked, something that happens most strongly perhaps in Arguedas's poem "Llamado a algunos doctores," which disputes occidental models of technological modernity by proposing an expanded definition of technology. Quijano equates Arguedas's commitment to translate "todas las posibilidades expresivas del idioma dominado," with "un programa de subversión lingüística," whose larger meaning would be that of "una propuesta de racionalidad alternativa" (Quijano 1988: 64). Yet the scene before the walls is less confident in its translation than is el Inca's book; it brings into evidence the possibility of an untranslatable residue. Is this, among other things, because the actually existing modernity of republican Peru—specifically that of Leguía's modernization, which is the period in which the novel is set—had failed, despite incorporating Inca motifs into the design of public buildings, to embody el Inca's utopia? Like Vargas Llosa's conversation, the contemplation of Inca stones is overdetermined.

The key passage of the novel is the following:

> Eran más grandes y extrañas de cuanto había imaginado las piedras del muro incaico; bullían bajo el segundo piso encalado que por el lado de la calle angosta, era ciego. Me acordé, entonces, de las canciones quechuas que repiten una frase patética constante: *"yawar mayu,"* río de sangre, *"yawar unu,"* agua sangrienta; *"puk'tik' yawar k'ocha,"* lago de sangre que hiere; *"yawar wek'e",* lágrimas de sangre. ¿Acaso no podría decirse *"yawar rumi,"* piedra de sangre, o *"puk'tik' yawar rumi,"* piedra de sangre hirviente? Era estático el muro, pero hervía por todas sus líneas y la superficie era cambiante, como la de los ríos en verano, que tienen una cima así, hacia el centro del caudal, que es la zona temible, la más poderosa. Los indios llaman *"yawar mayu"* a esos ríos turbios, porque muestran con el sol un brillo en movimiento, semejante al de la sangre. También llaman *"yawar mayu"* al tiempo violento de las danzas guerreras, al momento en que los bailarines luchan.
> -¡Puk'tik' yawar rumi!—exclamé frente al muro, en voz alta.
> Y como la calle seguía en silencio, repetí la frase varias veces. [13]

Over time, three ways of reading this passage have emerged. The first, which was developed furthest by Antonio Cornejo Polar, and which forms the basis for the others, emphasizes the role of translation: the meaning of the wall is carried across into the other culture through the translation of phrases drawn from Quechua songs: it is the living oral tradition which, brought into the genre of the novel and in the process transforming it,[14] enables the native culture to become available to the other. This process, for Cornejo, is necessarily utopian: in the carrying over of the living presence of the voice into writing there is an inevitable loss. The full bringing across of the voice into writing is "una utopía imposible." Thus he writes of "el canto andino," in the scene where Ernesto (the narrator-protagonist) imagines himself writing to the native (monolingual) girls he had known as a child, as a "modelo imposible" for writing. And yet he draws attention to the dramatic meaning of the moment when Ernesto speaks out aloud: "La traducción, que formalmente marca al quechua con itálicas y comillas, desaparece al final cuando la exultante exclamación de Ernesto borra esa ajenidad" (Cornejo Polar 1994: 214). Thus there is a tension that runs through Cornejo's reading, between translation and its limits, between utopia and loss, toward the latter of which he adopts a stance of pathos and perhaps melancholy.

The second way of reading this scene is the one put forward by Víctor Vich. Vich draws attention to "una cierta imposibilidad de traducción" of the native elements:

> En mi lectura, el muro se presenta como una amorfa masa de significado que Ernesto se esfuerza por descifrar pero que nunca se vuelve completamente inteligible. [. . .] En realidad, Ernesto toca el muro pero también se siente tocado por él: entre él y las piedras se establece un intercambio de significados que nunca conoceremos del todo bien. Se trata de un momento de radical extrañeza que lo convoca pero que al mismo tiempo lo repele. El muro interpela a Ernesto y tal llamado tiene como objetivo mostrarle la agónica densidad de la historia peruana. (Vich 2005: 368)

As Vich points out, this reading, with its acknowledgment of the untranslatable, would entail a shift of theoretical paradigms, away from that of transculturation to one of heterogeneity. The concept of transculturation, insofar as it proposes cultural translatability, would, therefore, be linked to that version of the temporality of the modern which holds that it is able to include the cultural and temporal other by making them readable.[15] Vich asks, "¿Qué es lo que la transculturación excluye para constituirse?" That prompts the question: Was the heterogeniety that Arguedas's scene makes evident already repressed by el Inca? Vich also asks, "¿Es posible imaginar la nación al margen de las prácticas homogenizadoras?" and points to unavoidable relationship between knowledge of the other culture and power: "la textualización de una cultura 'otra' había servido para dominarla con mayor productividad."

The third way of reading is the one that I have proposed in an essay that sets out to criticize the use of the dualism orality versus writing as a way of interpreting *Los ríos profundos*. I argue that this opposition makes invisible all those Andean textual practices, from *quipus* to weaving to ways of seeing the landscape,[16] that constitute forms of writing in an expanded, Derridean definition, and that constitute a range of cultural practices essential to the reproduction of native culture. In Arguedas's novel, those forms of inscriptive practice are concentrated into the device of the *zumbayllu*, which introduces into the ambit of occidental schooling native textual practices, that is, native ways of reading and writing.[17] In that connection, what makes the Inca past, for example, El Inca Roca, subsumed into the wall, readable, is knowledge of the Andean present, through the practice of folklore and ethnography, in particular through the study of native song in its ritual and historical contexts, something that Arguedas had been doing, and encouraging his pupils to do, from the mid-1930s. Thus when placed with the *zumbayllu*, that is, in the context of an expanded theory of reading, the scene before the Inca walls constitutes one of the most complex scenes of reading in Peruvian literature. It dramatizes what is meant by the cultural "other" in such a way as to make it impossible to draw a single line of distinction. And by placing on the scene a popular modernity, as opposed to the one produced and interpreted by the state, it complicates—perhaps refuses—the republican periodization of Peru and criticizes the temporal imagination which this periodization depends upon.

This third way of reading makes it impossible to place the other within a single time line or notion of development. That is the condition for its potential cancellation of Vich's lost otherness. Historical reality, in the sense of the forms of power that have dominated, has meant the opposite, that is, the imposition of a linearity called development. Thus the assertion of the third readability, the possibility of bringing it into reality, depended on the hope of social transformation that in *Yawar Fiesta* is conflictually ciphered in the name Mariátegui, and in *Los ríos profundos* in the triumphal entry of the *colonos* into Abancay. Readability becomes a question of political struggle or of what Alberto Flores Galindo called "forcing history."

History and Messianic Time in
Alberto Flores Galindo

Flores Galindo's book about Mariátegui, *La agonía de Mariátegui* (1980), is also a reflection on Peruvian temporality. One of the main claims it makes, in its reading of Mariátegui, is that his work made a radical break with the idea that the history of Peru would have to go through the same stages as the history of Europe, that is, from "Asiatic" forms of production, to feudalism, to capitalism, before socialism could become a possibility; in other words, that Mariátegui broke with the ideology of progress. But Flores also asserts an idea that is much harder to get to: the redemption of lived time by its projection into the future. César Vallejo's statement, "morir de vida y no de tiempo," and the desire it convokes, helps to open up this conception of time. It is difficult to grasp because it is implacably opposed to the standard time of modernity, "time cut up into equal abstract fragments" (Debord 1995: 107). Its conception of futurity is entirely different from that of uchrony, in that uchrony construes Peruvian history as a lack in comparison with the history of the "advanced" societies:[18] that is, it implies subjection to the model of an already-constituted continuum. Instead, what Flores finds in Mariátegui is an alternative sense of the task of the historian, "to blast open the continuum of history," as Walter Benjamin put it in 1940.[19] The Benjaminian conception of messianic time, where the present becomes "the 'time of the now' which is shot through with chips [i.e., fragments] of Messianic time" (Benjamin 1973: 265) helps to understand Flores's reading of Mariátegui. Time in this sense is absolutely non-sacrificial: it has nothing to do with renunciation in the present for the sake of some future transcendence. Nor has it to do with the accumulation of power within existing social relations. As Benjamin writes:

> The awareness that they are about to make the continuum of history explode is characteristic of the revolutionary classes at the moment of their action. The great revolution [i.e. the French Revolution] introduced a new calendar. The initial day of a calendar serves as a historical time-lapse camera. And, basically, it is the same day that keeps recurring in the guise of holidays, which are days of remembrance. Thus the calendars do not measure time as clocks do; they are monuments of a historical consciousness of which not the slightest trace has been apparent in Europe in the past hundred years. (Benjamin 1973: 263-4)

In a passage that illuminates the title he chose for his book, Flores quotes Mariátegui's statement, "La revolucíon más que una idea, es un sentimiento. Más que un concepto es una pasión," and comments: "En otras palabras, la Utopía "con mayúsculas," el mito, en cierta manera la religión de nuestro tiempo, la invitación a combatir por el milenio en la tierra: una agonía" (Flores Galindo 1994: 438-9).

La agonía de Mariátegui can be read, I would suggest, not just as a discussion of how to read Mariátegui 50 years after his death, but as a *mise en scène* of the work of the historian in the late twentieth century. Writing on Mariátegui, Flores locates his own practice as a historian, as author of *Apogeo y crisis de la república aristocrática* and *La utopía andina*, for example. In his introduction to the former, which traces the modernization of Peru between 1895 and 1930, he characterizes Peru as "un país múltiple" whose disparate parts are in process of becoming articulated through the growth of the internal market:

> sin embargo, a pesar de todos los cambios y renovaciones, el desarrollo del mercado interno es débil y los rasgos precapitalistas impregnan las relaciones personales y de trabajo, la mentalidad y las formas de existencia de los hombres de esta época. La heterogeneidad la hemos estudiado al tratar la hacienda azucarera, el latifundio andino y la fábrica limeña; los mecanismos de integración los hemos abordado a partir del comportamiento del capital mercantil en el sur andino. Así mostraremos cómo la República Aristocrática es una época de disparidades y conflictos entre lo nuevo y lo viejo. (Burga and Flores Galindo 1994: 26)

In other words, the method is to show heterogeneity and integration side by side, without any suppression of the real multiplicities. To think Peru as a totality only became possible, he shows, for that generation of intellectuals who migrated from the provinces to Lima in the second decade of the twentieth-century. Only then did it begin to be possible to imagine "una historia común": (Burga and Flores Galindo 1994: 269) "Para casi todos el Perú terminó siendo una esperanza, una utopía más que una realidad: concluyeron que el país no era una nación sino una posibilidad de nación, un 'concepto por crear' como decía Mariátegui en 1927" (Burga and Flores Galindo 1994: 266). In Flores's work, the deficit of uchrony is overcome by the capacity to think different times simultaneously.

But it is not a question of a confluence of different times. Confluence gives an image that is far too passive and natural. To think temporal heterogeneity requires an act of will. Mariátegui's sense of Peruvian time required an active bringing together of diverse sources: the archaeology of Inca civilization, the sociological investigation of the contemporary *comunidad indígena*, the study of peasant rebellions and participation in a peasant congress (Congreso de la Raza Indígena), Leguía's technological modernization (Flores Galindo 1994: 432–3). Out of his synthesis of these came his refusal of a linear, Eurocentrist image of history. That was after he had returned from Europe and experienced the effects of the Soviet Revolution in Italian politics. His earlier attitude had been more ironical: "Tenemos arte incaico. Teatro incaico. Música incaica. Y para que nada falte nos ha sobrevenido una revolución incaica" (Flores Galindo 1994: 537). His concept of Peruvian history placed him at odds with the linear program of the Communist International: the latter defined Peru as a semicolonial and feudal society, ready for a bourgeois-democratic revolution; for Mariátegui's Socialist Party, "la meta era, con absoluta claridad, una revolución socialista" (Flores Galindo 1994: 412). This meant ceasing to think given modes of production (native, feudal, capitalist) as components of a linear series that all nations were fated to fulfill, and bringing them together into a contemporaneous multi-temporality, which is itself part of a political practice, a struggle to transform them rather than a merely descriptive term. Flores comments: "Todo revolucionario así como busca insertarse en una tradición y formar parte de una historia para ejecutar una empresa colectiva, sabe que es igualmente necesario *forzar* a esa historia, actuar sobre el acontecimiento, llevar las posibilidades a sus límites" (Flores Galindo 1994: 511). He then notes that although showing the connections between Mariátegui and his time is necessary, it is not sufficient, that "el método histórico, a pesar de su aspiración a la totalidad" is itself insufficient.

Where might an image of that insufficiency be found? For Flores, it is in the idea of "agonía," which he presents in the epigraph to the book in the following statement by Mariátegui, written in 1924 and published in his book *El alma matinal*: "Agonía no es preludio de la muerte, no es conclusión de la vida. Agonía [. . .] quiere decir lucha. Agoniza aquel que vive luchando; luchando contra la vida misma. Y contra la muerte."

Epilogue: The Scene of Messianic Time

The conception of a zero time, in which temporal succession is broken, and present and future become one and the same thing, is, as Benjamin had noted, rarely embodied in actual historical experience. César Vallejo's Spanish Civil War book, *España, aparta de mí este cáliz*, perhaps gives the closest modern Peruvian approximation to that type of event. The desire for a zero time (or for what Mariátegui calls "la lucha final" in his essay of this title, collected in *El alma matinal*) has very little to do with the idea of foundational fictions and much more with an absolute cessation of occurrence. The poem "Masa," which is well known enough not to need quoting, is perhaps the strongest embodiment of that. That Vallejo found the image of messianic time in the Spanish Civil War does not make the temporality his book envisages any the less Peruvian. In his essay of 1937, "Los enunciados populares de la guerra española," Vallejo wrote: "Por primera vez una guerra cesa de ser una razón de Estado, para ser la expresión, directa e inmediata, del interés del pueblo y de su instinto histórico" (Vallejo 1997: 122). This constitutes, he states, "Un estado de gracia—así podríamos llamarlo—pocas veces dada a pueblo alguno en la historia" (Vallejo 1997: 124). This "enorme torbellino popular" is related "lo dionisíaco," which in Athenian democracy and subsequent forms of the state, had been expressly excluded from politics. It is a question of nothing less than "una nueva materia prima política"; thus the degree of change envisaged, which includes a new form of sovereignty: "La epopeya popular española [. . .] revela de cuánto es capaz un pueblo [. . .] Y todo este milagro—hay que insistir—lo consuma por obra propia suya de masa soberana, que basta a sí misma y a su incontrastable devenir" (Vallejo 1997: 125).[20] This is not the sovereignty of "lo popular," expressed through the political party that substitutes itself for the people, but a different type of sovereignty, expressed in the intelligence of the mass; in other words, an alternative modernity.

Notes

1. "Urge que el peruano sea cogido por sus preocupaciones y que luego por un mecanismo adecuado sea obligado a complicarse con otros peruanos en afanes más amplios, a luchar, a apasionarse, a acometer empresas, a exigir más, a ser responsable" (Basadre: 139). The awkward phrase "y que luego por un mecanismo adecuado" supplies the missing connection between past and future without telling us what it is. The lack of a concept allows common sense to fill the gap, with the idea of progress.

2. Basadre (1992: 151). See Antonio Cisneros's ironic poem on republican statues, "Descripción de plaza, monumento y alegorías en bronce," in Cisneros (1989: 54).

3. Palma compiled a dictionary of peruanismos which, when he took it to Spain, the Real Academia Española refused to accept. This sense of the need to modernize the language was shared by González Prada, who followed Andrés Bello's new system of spelling.

4. De la Fuente Benavides (1968: 349–50). *Deleznable* is an adjective that evokes the Lima of E. A. Westphalen, another Lima poet, in whose work, nevertheless, the logic of the visual is different (more of the twentieth-century), obeying a surrealist type of dream logic. Luis Rebaza points out: "The atmosphere of infantile evocations, decadence and disintegration in Eguren's poems, is an expression that accompanies a process of modernisation that has two faces: military destruction and the promise of technological progress. [. . .] Barranco [. . .] and

the neighbouring beach resort of Chorrillos were residential zones ransacked and burned by the occupying troops" (Rebaza 1997: 285–6).

5. Adán (1971: 59). This sentence is a fragment of the poem composed of fragments (written by one of the novel's characters, and called "*Poemas underwood*"), in which cinematic montage is placed in direct relation with the typewriter as technical machine that produces, through its capacity to tabulate, the transformation of sentences into paragraphs; i.e., the paragraph is the unit of composition, like the frame in cinema. Here two modern technologies work in synergy. (For the interconnection of spacing on the page and modern technologies of transport, see Mallarmé's *Un coup de dés;* for the use of paragraphs as units of composition that are also units of emotion, see Gertrude Stein, *How to Write.*) Note also: "El beso final ya suena en la sombra de la sala llena de candelas de cigarrillos. Pero ésta no es la escena final. Pero ello es por lo que el beso suena" (p. 59). The idea that you can hear the kiss because the film has not yet finished combines the aesthetics of silent film with the theme of aesthetic perfection as suicide. ("Nada me basta, ni siquiera la muerte; quiero medida, perfección, satisfacción deleite.")

6. Benjamin (1985). The history of photography was, as Benjamin points out, more complex than a simple juxtaposition of the old and the new can convey. Even in its very beginnings, traces of its future technical/aesthetic potential can be seen.

7. On cinematic substance, see Deleuze (1989: Ch. 2).

8. "este cinema [. . .] humoso." Adán, 53.

9. Gaonkar (1999: 154). "For instance, any culture could suffer the impact of growing scientific consciousness, any religion could undergo secularization, any set of ultimate ends could be challenged by a growth of instrumental thinking, any metaphysic could be dislocated by the split between fact and value. [. . .] These transformations may be facilitated by our having certain values and understandings, just as they are hampered by the dominance of others" (154–5).

10. Conrad (1929: 3). Conrad writes in the "Preface": "My task which I am trying to achieve is, by the power of the written word to make you hear, to make you feel—it is, before all, to make you *see*" (x).

11. His first novel, *La ciudad y los perros*, already included tracking shots and panoramic shots, but not the rhythmic use of cuts that characterizes *La casa verde*.

12. Arguedas's 1941 essay on Cusco, "El nuevo sentido histórico del Cusco" (in Arguedas, 1985: 131–8) can be taken as an intermediary text between *Comentarios reales* and *Los ríos profundos.*

13. Arguedas (1983: 14). I have italicized the Quechua phrases, in conformity with the first edition (Buenos Aires: Losada, 1958).

14. "La aguda tensión que genera la relación entre un instrumento cultural definidamente moderno y urbano, como era la novela, y una

instancia referencial (no sólo referencial [. . .]) que obedece a otras normas socio-culturales" (Cornejo Polar, 1994: 211).

15. See Peter Osborne's theorization of the time of modernity, where the emphasis is on the prior conceptual ground which that time provides and without which it would be impossible to bring other times into comparison. (Osborne 1995: 29).

16. I here draw on *El rincón de las cabezas* by Arnold and Dios Yapita (1998), which argues for Andean textual rights against neoliberal bilingual education. They show how the latter uses a narrow version of alphabetical writing and reading as a reverse mirror from which to define native Andean culture in Bolivia. In the process, the native theory of writing gets suppressed.

17. In Rowe (2003) I argue that in the novel these are carried over into alphabetic writing through an avant-gardist, ideographic, conception of the letter, which opposes the colonial (Augustinian) idea of the letter.

18. Chocano, "Ucronía y frustración," 50.

19. Benjamin (1973: 264). The full statement reads "man enough to blast open the continuum of history." The preceding sentence reads: "The historical materialist leaves it to others to be drained by the whore called 'once upon a time' in historicism's bordello."

20. The expression "nueva materia prima política" is from the essay "Los artistas ante la política," in Vallejo (2002: 517).

References

Adán, Martín (1971). *La casa de cartón*. Lima: Mejía Baca [1928].

Arguedas, José María (1983). *Los ríos profundos, Obras completas*, III. Lima: Horizonte.

—— (1985). "El nuevo sentido histórico del Cusco" [1941] in his *Indios, mestizos y señores*. Lima: Editorial Horizonte, pp. 131–8.

Arnold, Denise and Juan de Dios Yapita (1998). *El rincón de las cabezas*. La Paz: Ilca.

Basadre, Jorge (1992). *Perú: problema y posibilidad y otros ensayos*. Caracas: Biblioteca Ayacucho.

Benjamin, Walter (1973). "Theses on the Philosophy of History," in his *Illuminations*. Ed. Hannah Arendt. London: Fontana, pp. 255–66.

—— (1985). "A Small History of Photography," in his *One Way Street and Other Writings*. London: Verso, pp. 240–57.

Benjamin, William (2002). *The Arcades Project*. Cambridge, MA: Harvard University Press.

Burga, Manuel and Alberto Flores Galindo (1994). *Apogeo y crisis de la república aristocrática, Obras completas*, vol. II. Lima: Sur.

Chocano, Magdalena (1987). "Ucronía y frustración en la conciencia histórica peruana." *Márgenes* [Lima] I. 2: 43–60.

Cisneros, Antonio (1989). *Por la noche los gatos*. Mexico: Fondo de Cultura Económica.

Conrad, Joseph (1929). *The Nigger of the "Narcissus": A Tale of the Sea*. London: Dent.

Cornejo Polar, Antonio (1994). *Escribir en el aire: ensayo sobre la heterogeneidad socio-cultural en las literaturas andinas*. Lima: Editorial Horizonte.

De la Fuente Benavides, Rafael (1968) [Martín Adán]. *De lo barroco en el Perú*. Lima: Universidad Nacional Mayor de San Marcos.

Debord, Guy (1995). *The Society of the Spectacle*. New York: Zone Books.

Deleuze, Gilles (1989). *Cinema* II. London: Athlone.

Eguren, José María (1961). *Poesías completas*, comp. Estuardo Núñez. Lima: Universidad Nacional Mayor de San Marcos.

Flores Galindo, Alberto (1994). *La agonía de Mariátegui, Obras Completas*, Vol. II. Lima: Sur.

Gaonkar, Dilip (ed.) (1999). *Alter/Native Modernities*. Durham: Duke University Press.

González Prada, Manuel (1966). *Pájinas libres*. Lima: Fondo de Cultura Popular.

Martín-Barbero, Jesús (2002). "The Media: Memory, Loss and Oblivion." *GSC Quarterly* 4 (Spring). Available at http://www.ssrc.org/gsc/newsletter4/martinbarbero.htm

Osborne, Peter (1995). *The Politics of Time*. London:Verso.

Palma, Ricardo (2000). "Los ratones de fray Martín," in his *Tradiciones peruanas*. Madrid: Cátedra, pp. 565–8.

Quijano, Aníbal (1988). *Modernidad, identidad y utopía en América Latina*. Lima: Sociedad y Política Ediciones.

Rebaza, Luis (1997). "José María Eguren," in Verity Smith (ed.), *Encyclopedia of Latin American Literature*. London: Fitzroy Dearborn, pp. 285–6.

Rowe, William (2003). "Sobre la heterogeneidad de la letra en *Los ríos profundos*: una crítica a la oposición polar escritura/oralidad," in J. Higgins (ed.), *Heterogeneidad y literatura en el Perú*. Lima: Centro de Estudios Literarios Antonio Cornejo Polar, pp. 223–52.

Vallejo, César (1997). *Poesía completa*, vol. IV. Lima: Pontificia Universidad Católica del Perú.

——— (2002). "Los artistas ante la política," in his *Artículos y crónicas completos*, vol I. Lima: Pontificia Universidad Católica.

Vargas Llosa, Mario (1966). *La casa verde*. Barcelona: Seix Barral.

Vich, Víctor (2005). "El subalterno "no narrado": un apunte sobre la obra de José María Arguedas," in Carmen María Pinilla (ed.), *Arguedas y el Perú de hoy*. Lima: Sur.

Chapter 6

Belatedness as Critical Project: Machado de Assis and the Author as Plagiarist

João Cezar de Castro Rocha

Art and Society in Latin America: Crossing Roads

In a polemical essay, originally delivered as a seminar at a workshop on the role of the intellectual in exile, Joseph Brodsky attributed to himself the role of devil's advocate and, with the witty approach that distinguished his work, conveyed the toughness of a writer who resisted any self-indulgence regarding his personal circumstances:

> As we gather here, in this attractive and well-lit room, on this cold December evening, to discuss the plight of the writer in exile, let us pause for a minute and think of some of those who, quite naturally, didn't make it to this room. (. . .)
> Whatever the proper name for this phenomenon is, whatever the motives, origins, and destinations of these people are, whatever their impact on the societies which they abandon and to which they come, one thing is absolutely clear: they make it very difficult to talk with a straight face about the plight of the writer in exile.[1] (Brodsky 1994: 22–3)

The question of Latin America's troubled relationship with the concept of modernity justifies Brodsky's bitterness. After all, when discussing this question we might tend automatically to put ourselves outside the problem, a tendency that constitutes the blind spot of otherwise perceptive interpretations of Latin America's quest for modernity. It is as if we could see the problem from an external vantage point, instead of having our understanding determined by it. We should not speak of the "impasses" and "failures" of modernity in

Latin America without previous acknowledgment that we are not simply speaking *about* it, but *from* within an unfinished project. Therefore, I wonder whether this awareness demands a specific analytical perspective.

In this context, it is my contention that Machado de Assis was only able to create a groundbreaking work in the panorama of Western literature when he came to terms with the circumstance of Brazil as a "peripheral" country.[2] In Roberto Schwarz's apt definition, Machado was "a master on the periphery of capitalism" (Schwarz 2001). This particular location allowed him to develop what I would like to christen "belatedness as critical project." Nonetheless, I should clarify that I am not using the concept of belatedness to imply that a "peripheral" writer is always already coming or being after the expected time, which would be defined by the so-called central powers. Rather, in my reading of Machado de Assis, I am appropriating Jorge Luis Borges's "técnica del anacronismo deliberado y de las atribuciones erróneas."[3] Therefore, "belatedness as critical project" supposes a skeptical detachment from the hierarchy usually attributed to tradition as well as favoring an ironical gaze concerning contemporary values—trademarks of Machado de Assis's work. Moreover, I am aware of the pitfalls implied by any "triumphant interpretation of our backwardness."[4] However, if it is true that the concrete implications of such inequality should not be overlooked, at the same time, they should not predetermine the practice of literary criticism.

As a matter of fact, the issue of a belated modernity has haunted Latin American writers and social thinkers. In Brazilian cultural history, a matter of paramount importance is the question of and the quest for modernity, that is, economic progress, social justice, and, above all, the desire to be up-to-date with the latest trends of the so-called central powers. Brazilian cultural history, then, resembles a phantasmagorical race toward what has not yet been clearly indicated, and, therefore, cannot be fully achieved. In this context, however fast you travel, you always arrive after your prime. You are always already belated, especially if you run restlessly. For instance, this seems to be Charles Wagley's conclusion:

> Any book on Brazil should be published in 'loose leaf' form so that every few months certain pages might be extracted and rewritten. Brazil changes fast, events take sudden unexpected turns, and each year new articles and books appear on Brazilian society past and present. (Wagley 1971: vii)

Brazilian society seemingly defies interpretation, for it lacks even minimal stability in its ever-changing frenzy. If this observation is accurate, there is an embarrassing question to be addressed to the author, namely, why bother writing any book at all? Of course, the remark on the unheard-of velocity of changes is but a commonplace—hard to accept, especially when it comes from an anthropologist. However, in the late 1960s and early 1970s such a perception of the country was likely to be welcomed, although not for particularly good reasons. The military, who had seized power with the coup d'état of 1964, carefully construed and forcefully imposed the image of an unparalleled rate of development during the so-called Brazilian miracle, which was supposed to modernize the country as well as redistribute the wealth that was to be created. The complete failure of this policy is well known and I need not dwell on it. Let me rather bring to this discussion Marshall Eakin's alternative interpretation of the pace of structural transformation in Brazil.

All nations carry with them the scars of their past, yet Brazil bears the "burden of history" more visibly than most. Everywhere one looks in Brazil, the past intrudes upon the present. The modern, rapidly changing "country of the future" appears unable to escape another traditional and unchanging Brazil that is seemingly frozen in time.[5]

Eakin adds a nuance, which is instrumental to my reading of Machado de Assis. Instead of defining Brazil as the country of the future or, inversely, as if the future itself has always already been there, Eakin stresses the challenge posited by Latin American cultural history, namely, the need to cope with the simultaneity of different historical times—and ultimately with their clash. As I will argue, authors such as Machado de Assis were able to transform this complexity into formal challenges to the tradition of the novel developed in modern times.

Euclides da Cunha registered such a clash between contradictory perceptions of historical times with vigor in his masterpiece, *Os Sertões*. His book still stands today as a vehement admonition on the complexity of this circumstance, and, above all, a warning of the consequences of the imposition of one viewpoint at the expense of other alternatives. Let us remember a quotation extracted from *Rebellion in the Backlands*' "Preliminary Note." However, let us not forget to place in context Cunha's vocabulary concerning "great" or "backward" races. This hermeneutic procedure is needed in order to focus on the important issue touched upon by Cunha. His book is ultimately a powerful reflection on the results of the overlapping of conflicting perceptions of time:

> The first effects of various ethnic crossings are, it may be, initially adapted to the formation of a great race; there is lacking, however, a state of rest and equilibrium, which the acquired velocity of the march of the peoples in this century no longer permits. Backward races today, tomorrow these types will be wholly extinguished. (Cunha 1944: xxxi)

Undoubtedly, Cunha was imbued with the prejudices of his time. In spite of that, he displays a keen understanding of the main dilemma of Brazilian society: the copresence of different perceptions of time as well as the historical impossibility of reconciling them. The war of Canudos, fought from November 1896 to October 1897, was a blatant symptom that the process of modernization did little to acknowledge the country's disparities. More than a century later, is it not true that there is a similar problem in the shantytowns [favelas] of Rio de Janeiro? Is it not even more disturbing to learn that the very word "favela" was incorporated into Brazilian social vocabulary as an outcome of the civil war so vividly described in Cunha's *Os Sertões*?[6] Once more, we face the dialectics of a structure seemingly unchanged amidst the ceaseless flow of high-speed transformations.

Therefore, a question imposes itself: To what extent does this dilemma concern the relationship between art and society in Latin America? I propose that two sides of this dilemma can be discerned.

First of all, the social consequences of the simultaneity of historical times have almost always been antagonistic, for instead of any effort to understand what has prevailed there has been the violent imposition of one view upon another. This was most starkly evident in the eloquent but threatening formula put forward by Sarmiento's uncompromising statement: "civilización y barbarie,"[7] which easily translates itself as "civilización *o* barbarie." As the sheer dualism of the sentence implies, there is no room for accommodating differences—modernity itself becomes a value that has to be fully accepted, in spite of the irreversible changes entailed in its implementation. Euclides da Cunha quoted Sarmiento's essay with due respect, although naming it *Civilización y barbarie*, quietly changing the subtitle into its title![8] In this case, more than a Freudian slip, the misquotation reveals Cunha's synthesis of Sarmiento's book. As a matter of fact, Cunha rewrote the sentence in *Os sertões*, giving a dramatic turn to the formula, adding a fatalistic undertone to Sarmiento's resolution: "We are condemned to civilization. Either we shall progress or we shall perish. So much is certain, and our choice is clear."[9] Therefore, the uneasy nature of the dilemma brought about by this uncanny overlapping of temporalities comes to the fore. Indeed, this overlapping, the disquieting consequences of which have to be reckoned with, characterizes most of Latin America's social fabric.

On the other hand, the same constellation of problems might produce a radically different artistic outcome. My reading of authors such as Machado de Assis relies on this hypothesis. As far as artistic production is concerned, the overlapping of historical times might be particularly inspiring, if not propitious to the development of what I proposed to call "belatedness as critical project."

A Belated Writer—Ahead of his Time

The work of Machado de Assis will be analyzed here to illustrate this hypothesis. His first important and innovative novel, *The Posthumous Memoirs of Brás Cubas*, was published in serial form in the *Revista Brasileira*, in 1880, and in book form the next year. This groundbreaking work has been praised as a masterpiece by writers and critics such as Carlos Fuentes, Susan Sontag, John Barth and Harold Bloom, among others—let alone Brazilian readers. Nonetheless, until the writing of *Brás Cubas*, although he was already a noted author, respected by his contemporaries, Machado de Assis's work did not yet have the edginess that we have learned to admire. Therefore, one of the most pressing questions of Brazilian literary criticism is the need to provide a plausible explanation for the authentic quantum leap represented by Machado's works after 1880.[10] In this essay, I will naturally not provide a summary of the state of the art on this topic. I will instead invite the reader to open the first page of the novel. She will find a note "To the Reader":

> That Stendhal should have confessed to have written one of his books for a hundred readers is something that brings on wonder and concern. Something that will not cause wonder and probably no concern is whether this other book will have Stendhal's hundred readers, or fifty, or twenty, or even ten. Ten? Five, perhaps. The truth is that it's a question of a scattered work where I, Brás Cubas, have adopted the freeform of a Sterne or a Xavier de Maistre. I am not sure but I may have put a few fretful touches of pessimism into it. It's possible. The work of a dead man. I wrote it with a playful pen and melancholy ink (. . .).[11]

This is a key passage; indeed, this is Machado de Assis's rite of passage. After the very beginning of *The Posthumous Memoirs of Brás*

Cubas the narrator fashions himself as an author who fully acknowledges that, above all, he is a *reader*, a statement that undermines Romantic notions of authorship.[12] As we will see, it is clear that within this conception Harold Bloom's theory of "the anxiety of influence" reveals itself as a Romantic projection of the notion of "genius," which is precisely what is called into question by authors such as Machado de Assis. As of 1880, the surface of his texts is ever more populated by innumerable references to authors, topics, and tropes of the literary tradition. If Machado consciously assimilates Sterne's technique of digression, he does so with Montaigne's flavor, so to speak, for his digressions usually start or end with literary references. As Alfred Mac Adam has noted: "Through this reference to *De l'amour*, which blurs essay, fiction, and poetry, and his later references to Sterne and Xavier de Maistre, Brás creates antecedents for his disconnected *Posthumous Memoirs*."[13] Had Jorge Luis Borges read Machado's novel, then the Argentinean could have written a new essay—"Machado and his Precursors." And it should be noted that Brás Cubas is a "deceased author," who starts his career in a special circumstance: after his death.[14] This uncanny "delegation of the writing to the dead man displaces the fiction toward an intransitive and artificial authorial freedom" (Hansen 1999: 42), which should engage the reader into a renewed fictional pact. In other words, Machado not only fashions himself as a reader, but he also impels the reader of his novels to acknowledge her role in the constitution of the fictional play.

As a matter of fact, since his first novel, *Ressurreição*, published in 1872, Machado portrayed himself as a worker, who was determined to craft his skills in a genre in which he was a beginner.[15] In *Posthumous Memoirs*, especially in the note "To the Reader," Machado takes a step further. He not only renders explicit the authors with whom he is dialoguing, but also provides a conceptual framework to this dialogue: he is interested in the "free-form."[16] Moreover, he imposes upon this form a particular twist. Machado will not digress endlessly or travel around his chamber, propelled by witty humor as the pilot of his journey. As he states clearly, his itinerary will demand a copilot, that is, to the "playful pen," he will add a "melancholy ink." Machado, therefore, brings together the eighteenth and the nineteenth centuries in the figures of Sterne, Xavier de Maistre and Stendhal as well as introducing humor to the somber mood of melancholy. Machado had already envisioned the technique of the "deliberate anachronism" in this overlapping of historical times as well as of literary genres. Modernity as an unfinished process is not necessarily experienced as an impasse, but rather as an opportunity to encompass simultaneously different horizons. Thus, Machado is not nostalgic for an idealized view of Brazilian history, to be preserved against the process of modernization. At the same time, he is not enthusiastic about the promises of modernity. In other words, the free-form of his prose corresponds to the free-form of his thinking.

This explicit acknowledgment of the simultaneity of different historical times produced an awareness that distinguishes Machado's achievements. It is as if a peripheral writer has to face a phenomenon that could be called "compression of historical times," namely, she receives simultaneously information from several historical periods, without the "benefit" of a reasonable chronological order or an already stable interpretive framework. In Brazilian literature this problem has always already been there; after all, "the novel has existed in Brazil before there were Brazilian novelists. So when they appeared, it was natural that they should follow the European models, both good and bad, which had already become entrenched in *our reading habits*."[17] The usual answer to this situation is the development of an "anxiety of up-to-dateness," which obliges the writer to engage in an impossible race, for there will never be an adequate starting point— wherever you begin, there is already previous ground to be covered. Carlos Fuentes has humorously targeted such anxiety: "Las imitaciones extralógicas de la era independiente creyeron en una civilización Nescafé: podíamos ser instantáneamente modernos excluyendo el pasado, negando la tradición" (Fuentes 2001: 10).

Nonetheless, there is an alternative, exercised by an author such as Machado de Assis,[18] for whom the clash of historical perceptions becomes a literary device of unique strength. This device renders productive, at the formal level, the historical precedence of *reading* over *writing*. In other words, Machado brings to the structure of his composition the fact that, in Latin America and not only in Brazil, "the novel has existed before there were novelists." Therefore, in Latin America the first novelists were the attentive and sometimes critical readers of European novels. It is true, however, that, to a degree, this circumstance applies to all literatures—this acknowledgment is indispensable, in order to avoid another naive eulogy of belatedness. In the case of Latin America, where the colonial past was recent, the prevalence of the act of reading produced a predictable and collective "anxiety of influence." On the contrary, toward the end of the century, Machado was able to welcome the notion of a fundamental lack of originality, which becomes a liberating force. If there is no possibility of fashioning oneself as an "original" writer, then, the whole of literary tradition might be freely appropriated. Thus, Machado's conflation of several centuries of literary tradition, literary genres, and, above all, the acts of reading and writing fully announced Borges's "anacronismo deliberado." In an acute reading of the Brazilian author, Carlos Fuentes remarked:

> Y sin embargo, el hambre latinoamericana, el afán de abarcarlo todo, de apropiarse todas las tradiciones, todas las culturas, incluso todas las aberraciones; el afán utópico de crear un cielo Nuevo en el que todos los espacios y todos los tiempos sean simultáneos, aparece brillantemente en las *Memorias póstumas de Blas Cubas* como una visión sorprendente del primer Aleph, anterior al muy famoso de Borges (. . .). (Fuentes 2001: 24)

Therefore, Machado was able to transform the notion of belatedness, which accompanies the process of peripheral modernization,[19] into a critical project. Is it not true that, at the time of the prevalence of the French school of comparativism, a "peripheral" author was commonly interpreted as the outcome of the "influences" received from metropolitan writers? If so, Machado seems to ponder: allow this author to become at once a malicious reader, an imaginative writer, and, above all, a skeptical critic regarding hierarchies and literary glories. It is as if Machado knew that the question of the international repercussions of his work was at once unavoidable and irrelevant.[20] On the one hand, it is unavoidable; after all, peripheral countries keep on searching for legitimacy, which comes from abroad. Moreover, this question is ultimately irrelevant, for such legitimacy implies only that peripheral authors have satisfied the exotic expectations imposed upon their culture.[21] In that case, they would have indulged in the regrettably common phenomenon of "self-exoticism," as Jorge Amado's late work illustrates.

Machado's undermining of traditional notions of authorship also expresses his divergence from established views of his time. His insightful answer to the problem of literary modernity in Latin America, through the questioning of the acts of reading and writing, was taken further in his next novel, *Quincas Borba*, published in 1891. In chapter CXIII, the reader is introduced to the following situation: Rubião, the faithful although unwise follower of the philosopher Quincas Borba, inherits the fortune of his master, and starts spending it recklessly. One of his enterprises is the funding of a political newspaper, whose owner—Camacho, an unscrupulous lawyer and journalist—is only interested in benefiting from Rubião's naïveté. One day, Rubião visits the newsroom and casually reads an article. Even more randomly, he suggests minor changes to its composition. Naturally

Camacho adopts his patron's emendations. Rubião is delighted, and, through a humorous chain of associations, decides that he is the true author of the entire piece. In Machado's words, Rubião's reaction could provide the title to a new chapter: ' "How Rubião, satisfied with the correction made in the article, composed and pondered so many phrases that he ended up writing all the books he'd ever read.' "[22] There is, of course, a logical problem in this uncannily fast transition from reading books to composing all of them. Machado offers a solution:

> There is a gap between the first phrase saying that Rubião was co-author and the authorship of all books read by him. What certainly would be the most difficult would be going from that phrase to the first book—from there on the course would be rapid. It's not important. Even so, the analysis would be long and tedious. The best thing is to leave it this way: For a few moments Rubião felt he was the author of many works by other people.[23]

This passage is akin to the spirit of the most celebrated short stories by Jorge Luis Borges, especially the ones devoted to issues of readership and authorship. As Silviano Santiago has insightfully remarked, based on an innovative reading of "Pierre Menard, autor de Quijote": "(. . .) the Latin American writer is a devourer of books. He reads constantly and publishes occasionally."[24] If we follow Rubião's method, we will then understand that the Latin American writer does not publish more often because there is no volume that potentially was not written by her hungry eyes.

In Machado's next novel, *Dom Casmurro*, published in 1899, the question of authorship is once more of paramount importance. For instance, Bento Santiago, the first-person narrator, clarifies that the title of the novel was due to an unfortunate incident. One day, returning home in a train, he met a neighbor, a young man; indeed, a "poet" who decides to recite his complete works. Naturally, the narrator falls sleep, infuriating the unknown "genius."[25] As revenge, he decides to nickname his inconsiderate neighbor, and chooses to call him "Casmurro." The narrator elucidates the epithet: "(. . .) the [meaning] the common people give it, of a quiet person who keeps himself to himself."[26] Or: someone who is not polite enough to undergo some minutes of embarrassing poetry. "Dom" was added as a sign of mockery, since Bento Santiago certainly did not have an aristocratic life. However, instead of being upset, the narrator transforms the nickname into the title of his memoirs: *Dom Casmurro*. He even bestows on the young poet an unexpected possibility:

> Still, I couldn't find a better title for my narrative; if I can't find another
> before I finish the book, I'll keep this one. My poet on the train will
> find out that I bear him no ill will. And with a little effort, since the title
> is his, he can think that the whole work is. There are books that only
> owe that to their authors: some not even that much.[27]

Machado de Assis as Reader/Readers of Machado de Assis

Therefore, Machado de Assis affirmed his uniqueness through the role of a reflective reader who eventually becomes a self-reflective author, whose text is primarily the written memory of his private library. It is now time to turn to some critics and authors in order to see how they reacted to Machado's innovations: from Machado as reader to the readers of Machado.

In his late exuberant style, Harold Bloom considered Machado de Assis "one of the hundred exemplary creative minds" of Western literature. However, the best explanation for his conclusion is the obvious, but not always properly understood relationship between the author of *Dom Casmurro* and the author of *Tristram Shandy*:

> This is not to deny originality and creative zest to the Brazilian master, but
> only to remark that Sterne's spirit freed Machado from any merely nation-
> alistic demands that his Brazil might have hoped to impose upon him.
>   Machado de Assis is a kind of miracle, another demonstration of the
> autonomy of literary genius in regard to time and place (. . .). (Bloom
> 2002: 675)

First of all, it is remarkable that Bloom does not really emphasize Machado's constant allusions to and rewritings of Shakespeare's works. No other author was so important to the reader of Machado de Assis. *Dom Casmurro* is a radical reading, that is, rewriting of *Othello*. Helen Caldwell has examined the case in her groundbreaking *The Brazilian Othello of Machado de Assis: A Study of Dom Casmurro*. As a matter of fact, Machado was obsessed with this particular play: "Shakespeare's *Othello* is brought into the argument of twenty-eight stories, plays, and articles" (Caldwell 1960: 1). Machado's rewriting brings to the fore a potential contradiction. Is it not true that, for all the force of Iago's malice, it was Othello's insecurities regarding the position he occupied that allowed Iago's intrigues to work on him? Machado creates an Othello who is at the same time his own Iago.

Thus, Othello's drama is reenacted, but with the suppression of the character of Iago. This clever artifice renders even clearer the nature of jealousy, portrayed as a feedback system, which, regardless of objective evidence, feeds on itself.[28] Bento Santiago, the first-person narrator of the novel, takes more than 200 pages to convince the reader (and, above all, himself) that his wife, Capitu, betrayed him with Escobar, allegedly his best friend. And the more he tries to present his case before the jury, that is, before the readers, the less he seems to persuade them—without an Iago to blame, how is it possible to explain an apparently uncalled-for jealousy, except by doubting the jealous person instead of his or her partner? According to one perceptive critic, the novel stages "a parody of tragedy, a systematic falsifying of all evidence, the text is a literature on literature, a fiction on fiction" (Hansen 1999: 43). Moreover, Machado's rewriting is literally a reflection on the intertwinement between the acts of reading and writing.

Machado offers yet another beautiful homage to Shakespeare, which once more highlights his thoughtful undermining of traditional concepts of authorship. In a chapter properly entitled "The Opera," the narrator remembers the curious theory of an old Italian tenor, according to whom the world was neither a dream nor a stage, but an opera. Literally so—Marcolino explains: "God is the poet. The music is by Satan. (. . .)."[29] After his expulsion from Heaven, Satan stole the manuscript from the Heavenly Father, and composed the score, which, at first, God did not want to listen to. Upon Satan's insistence, He decides to stage the opera, creating "a special theatre, this planet, and invented a whole company."[30] Some paragraphs later, the reader finds the corollary to Marcolino's theory:

> The element of the grotesque, for example, is not to be found in the poet's text: it is an excrescence, put there to imitate *The Merry Wives of Windsor*. This point is contested by the Satanists, with every appearance of reason. They say that, at the time when the young Satan composed his opera, neither Shakespeare nor his farce had been born. They go as far as to affirm that the English poet's only genius was to transcribe the words of the opera, with skill and so faithfully that he seems to be the author of the composition; but of course he is a plagiarist.[31]

This perhaps sounds an odd eulogy. After all, how do we consider that an author excels in his creation exactly when he allows himself to become an original plagiarist? The paradox seems unavoidable, but only if one holds Romantic notions of authorship, in which the "anxiety of influence" is as contagious as Othello's and Bento Santiago's jealousy. However, if a writer envisions her own location as precarious, then, the acknowledgment of previous "influences" (and let us use the term in order to dialogue with Bloom's theory) cannot be experienced as anxiety; rather they become liberating, for being influenced opens up the doors of the literary tradition as a whole. Caldwell perfectly understood Machado's appropriation of literary tradition: "The best way of comprehending the universal soul of mankind, said Machado, was through study of great writers the world over; the best way of portraying it was by 'plagiarizing' them."[32] Among others, Enylton de Sá Rego has shown the amplitude of Machado's readings, underscoring his affiliation to Menippean Satire.[33] Machado rendered clear that a creative author is above all a malicious reader of the tradition, which then becomes a vast and tempting menu, whose list of options is to be appreciatively savored and, to use a word that Machado was particularly fond of, ruminated upon as many times as needed for a proper digestion, that is, the composition of the next book. Once more, this is the literary device that transforms belatedness into a critical project. After all, Machado himself explained the "difference between literal quotations—which simply invoke someone else's authority—and the really artistic quotations— which creatively rewrite the quoted authors" (Sá Rego 1997: xvii). Thus, there is hardly any higher praise than considering an author to be authentic metonymy of plagiarism—Shakespeare.

Bloom missed another point. Machado did not excel as an author in spite of his time and place, but, rather, he could develop a highly original approach to the notions of authorship and readership pre-cisely because, as we have seen, he was "a master on the periphery of capitalism," in the sharp definition proposed by Roberto Schwarz, one of the most innovative readers of Machado de Assis. John Gledson has provided the best synthesis of Schwarz's theory:

> The great achievement of *A Master*, I think, is to explain an apparent paradox: how is it that a writer so rooted in his own time, writing in a slave-owning cultural backwater, is also, in many ways, so *advanced*? Schwarz's great perception (. . .) is that the modernity paradoxically arises, to a considerable degree, out of the backwardness, and does not merely happen in spite of it.[34]

Moreover, precisely by not being located at the center of the capitalist world in his provincial Rio de Janeiro, in the last decades of the nineteenth century, Machado was able to direct an especially keen critical gaze upon notions that were imposed as universal. The parody of scientific theories of the age, embodied in what he called "Humanitism," is a perfect illustration of a sophisticated mockery of Positivism, Social Evolutionism, Behavioral Psychology, and even Spiritism. In Chapter CXVII of *The Posthumous Memoirs of Brás Cubas*, there is an overt parody of Comte's philosophical system, focused on the arbitrary establishment of three phases throughout the course of mankind's history: "(. . .) Humanitas has three phases: the *static*, previous to all creation; the *expansive*, the beginning of things; the *dispersive*, the appearance of man; and it will have one more, the *contractive*, the absorption of man and things."[35] The three moments suddenly are transformed in four steps—after all, why not two phases or five periods? The incoherence, disguised under the rationale of a scientific discourse, is brought to the fore by Machado's fictional derision.[36]

This witty disposition associated with a skeptical view of human nature justifies John Barth's interest in the Brazilian author. The following quote, indeed, is one of the most acute definitions of Machado's achievements:

(. . .) I discovered by happy accident the turn-of-the-century Brazilian novelist Joaquim Maria Machado de Assis. Machado—himself much under the influence of Laurence Sterne's *Tristram Shandy*—taught me something I had not quite learned from Joyce's *Ulysses* and would not likely have learned from Sterne directly, had I happened to have read him: how to combine formal sportiveness with genuine sentiment as well as a fair degree of realism. Sterne is Pre-Romantic; Joyce is late or Post-Romantic; Machado is both Romantic and romantic: playful, wistful, pessimistic, intellectually exuberant. He was also, like myself, a provincial (. . .). (Barth 1989: vi–vii)

A provincial is a plagiarist by the very location of her culture. Her gesture of reproducing other cultures always implies, at least potentially, the gesture of mockery, the attitude of critical detachment. Moreover, Barth conflates in Machado's work two opposing historical perceptions: Machado would be "Pre" as well as "Post" whatever concept one attaches to his fiction. Once more, the "playful pen" and the "melancholy ink" come to the fore. The strength to be derived from the simultaneous perception of contradictory viewpoints was also stressed by Susan Sontag:

Our standards of modernity are a system of flattering illusions, which permit us selectively to colonize the past, as are our ideas of what is provincial, which permit some parts of the world to condescend to all the rest. Being dead may stand for a point of view that cannot be accused of being provincial. *The Posthumous Memoirs of Brás Cubas* is one of the most entertainingly unprovincial books ever written. And to love this book is to become less provincial about literature, about literature's possibilities, oneself. (Sontag 2002: 39–40)

Peripheral, provincial: different names to voice what Machado really was—a creative reader; a plagiarist. I may then conclude proposing another definition of the plagiarist. He is an author who "refuses to accept the traditional notion of artistic invention since he himself denies the total freedom of the artist."[37] He is a writer whose originality is his awareness that no author should *desire* to be portrayed as "original." After all, an "original" writer is someone who ultimately is not sufficiently well read or whose library only contains uninteresting volumes. If it is true that there are authors who publish more than they write,[38] inversely, the plagiarist is an author who has read much more than he could ever publish. The reader already knows where I am heading: Machado de Assis is one of the first authors in the tradition of Western literature to have been fully aware that he was first and above all a *reader*. Jorge Luis Borges has already christened the plagiarist who becomes a great author. His name is Pierre Menard. However—and in spite of the fact that, as Susan Sontag guessed, "Borges, the other supremely great writer produced on that continent, seems to have never read Machado de Assis" (Sontag 2002: 39)—Borges would not disagree if different names were attributed to the plagiarist. According to the old Italian tenor: Shakespeare. Or, an obsessive reader of *Othello*: Machado de Assis.

Notes

1. I owe this reference to Henning Ritter.
2. I am using the concept of "peripheral" not as an objective description of a given place, but rather as a complex of asymmetrical political, economic, and cultural relationships, "peripheral" being the pole located in a hierarchically secondary position.
3. This is the context of the famous passage: "(. . .) la técnica del anacronismo deliberado y de las atribuciones erróneas. (. . .) Esa técnica puebla de aventuras los libros más calmosos. Atribuir a Louis Ferdinand Céline o a James Joyce la *Imitación de Cristo* ¿no es una suficiente renovación de esos tenues avisos espirituales?" (Borges (1989: 450).
4. Roberto Schwarz, "Brazilian Culture: Nationalism by Elimination," in Schwarz (1992: 7). Some paragraphs earlier, the argument was made even clearer, through a remark on Foucault's and Derrida's work: "One

can easily appreciate how this would enhance the self-esteem and relieve the anxiety of the underdeveloped world, which is seen as a tributary to the central countries. We would pass from being a backward to an advanced part of the world, from a deviation to a paradigm, from inferior to superior lands (although the analysis set out to surpass just such superiority)." *Idem*, 6.

5. Eakin (1998: 7). The title of Eakin's book toys with the title of Stefan Zweig's 1941 *Brazil: Land of the Future.*

6. I have addressed these issues in Castro Rocha (2005).

7. Of course, I am referring to Domingo Faustino Sarmiento's *Vida de Juan Facundo Quiroga. Civilización y barbarie.*

8. I owe the reference to Cunha's misquotation to Zilly (2001: 104).

9. Cunha (1944: 54). As far as the deterministic note is concerned, it should suffice to read what follows on from the quotation: "This is scarcely suggested, it may be, by the heterogeneity of our ancestral element; but they are reinforced by another element equally ponderable: a physical milieu that is wide and varied and, added to this, varied historical situations which in large part flow from that milieu." *Idem*, 54.

10. Roberto Schwarz has perfectly formulated this problem: "The discontinuity between the *Posthumous Memoirs* and the somewhat colorless fiction of Machado's first phase is undeniable, unless we wish to ignore the facts of quality, which are after all the very reason for the existence of literary criticism. However, there is also a strict continuity, which is, moreover, difficult to establish" (2001: 149).

11. Assis, "To the Reader," (1997: 5). The original (1977b: 97) reads: "Que Stendhal confessasse haver escrito um de seus livros para cem leitores, é coisa que admira e consterna. O que não admira, nem provavelmente consternará é se este outro livro não tiver os cem leitores de Stendhal, nem cincoenta, nem vinte, e quando muito, dez. Dez? Talvez cinco. Trata-se, na verdade, de uma obra difusa, na qual eu, Brás Cubas, se adotei a forma livre de Um Sterne ou de um Xavier de Maistre, não sei se lhe meti algumas rabugens de pessimismo. Pode ser. Obra de finado. Escrevia-a com a pena da galhofa e a tinta da melancolia (. . .)."

12. Bluma Waddington Vilar has proposed an insightful reading of this problem in her Ph.D. dissertation (Vilar 2001). See especially the chapter, "Citação e autobiografia: *Memórias póstumas de Brás Cubas*," 118–51. Vilar combined Machado's undermining of traditional notions of authorship with a careful study of what she calls the "Machado de Assis system of citation."

13. "Review." Alfred Mac Adam. *Hispanic Review*, Winter (2000) 68: 97.

14. As Brás Cubas explains to the reader: "(. . .) I am not exactly a writer who is dead but a dead man who is writer, for whom the grave was a second cradle (. . .)." Assis, "To the Reader" (1997: 7). The original reads: "(. . .) é que eu não sou exatamente um autor defunto, mas um defunto autor, para quem a campa foi outro berço (. . .)" (1977b: 99).

15. "Already in the "Warning to the Reader," placed at the beginning of *Ressurreição* [Resurrection], after introducing himself to the critics as a "worker," (. . .) he concedes all creative power to "reflection" and "study." He finally rejects for himself the condition and law of genius (. . .)." Santiago (2001: 65).
16. Sergio Paulo Rouanet (2005) is currently developing an important reading of the relationship between Machado de Assis and the authors quoted in *The Posthumous Memoirs of Brás Cubas*.
17. Roberto Schwarz. "The Importing of the Novel to Brazil and its Contradictions in the Work of Alenca," in Schwarz (1992: 41). My emphasis.
18. Indeed, following on from the passage just quoted, Fuentes concludes (2001: 10): "El genio de Machado se basa, exactamente, en lo contrario: su obra está permeada de una convicción: no hay creación sin tradición que la nutra, como no habrá tradición sin creación que la renueve."
19. For the concept of "peripheral modernity," see Sarlo (1988).
20. It is likely that Machado would have read the following passage with an ironic (although self-contained) smile: "Machado de Assis is no longer unknown among us. Four of his novels and some fifteen or so short stories have now appeared in English and have been greeted with a kind of indignant wonder that this Brazilian author who was born in 1839 and died in 1908 was not even a name to us" (Caldwell 1970: 3).
21. Although important, this discussion would deviate from the main purpose of this chapter. Nonetheless, let me recommend Earl Fitz's analysis of the problem (1989). Fitz asks the question: "Why has it taken so long for Machado to begin to receive the international acclaim he deserves?" And he provides the answer: "Portuguese (. . .) is simply not widely recognized as a literary language in which quality literature is written." Therefore, "the truth, unfortunately, is that Brazilian literature is not recognized as constituting a significant part of Western literature. *Idem*, 10–1.
22. Assis (1998, CXIII: 160). The original (1975: 245) reads: "De como o Rubião, satisfeito da emenda feita no artigo, tantas frases compôs e ruminou, que acabou por escrever todos os livros que lera. (. . .)."
23. Assis (1998, CXIII: 160). The original (1975: 246) reads: "Há um abismo entre a primeira frase de que Rubião era co-autor até a autoria de todas as obras lidas por ele; é certo que o mais que mais lhe custou foi ir da frase ao primeiro livro;—deste em diante a carreira fez-se rápida. Não importa, a análise seria ainda assim longa e fastidiosa. O melhor de tudo é deixar só isto; durante alguns minutos, Rubião se teve por autor de muitas obras alheias."
24. Silviano Santiago, "Latin American Discourse: The Space In-Between," in Santiago (2001: 37).

25. "The journey was short, and it may be that the verses were not entirely bad. But it so happened that I was tired, and closed my eyes three or four times; enough for him to interrupt the reading and put his poems back in his pocket." Assis (1997). I—The Title, 3. The original (1977a: 67) reads: "A viagem era curta, e os versos pode ser que não fossem inteiramente maus. Sucedeu, porém, que como eu estava cansado, fechei os olhos três ou quatro vezes; tanto bastou para que ele interrompesse a leitura e metesse os versos no bolso."

26. Assis (1997). I—The Title, 4. The original (1977a, 67) reads: "(. . .) mas no [sentido] que lhe pôs o vulgo de homem calado e metido consigo." Helen Caldwell (1960, 2) mistrusts the narrator's elucidation, and asks keenly: "The definition he did not want us to see is this: 'an obstinate, moodily, stubborn, *wrong-headed* man.' Perhaps we will decide that this older definition fits Santiago better than the one *he* offers."

27. Assis (1997). I—The Title, 4. The original (1977a: 67) reads: "Também não achei melhor título para a minha narração; se não tiver outro daqui até ao fim do livro, vai este mesmo. O meu poeta do trem ficará sabendo que não lhe guardo rancor. E com pequeno esforço, sendo o título seu, poderá cuidar que a obra é sua. Há livros que apenas terão isso dos seus autores; alguns nem tanto."

28. Caldwell (1960: 1) makes an insightful remark about this issue: "Jealousy never ceased to fascinate Machado de Assis. (. . .) Jealousy has a fat part in seven of his nine novels; the plots of ten short stories turn upon the ugly passion—though in seven of the latter, to be sure, it receives an ironic if not rudely comic treatment." Silviano Santiago has also stressed this factor in Machado's fiction, explaining "(. . .) how the problem of jealousy arose in the Machadian universe. It comes (. . .) from the character's conception of the nature of love and marriage, as well as, on the other hand, the delicate games of *marivaudage* that man and woman have to represent to be able to arrive at union." "The Rhetoric of Verisimilitude," in Santiago (2001: 66). See also Param (1970: 198–206).

29. Assis (1997). IX—The Opera, 18. The original (1977a: 78) reads: "Deus é o poeta. A música é de Satanás (. . .)."

30. Assis (1997). IX—The Opera, 19. The original (1977a: 78) reads: "Criou um teatro especial, este planeta, e inventou uma companhia inteira (. . .)."

31. Assis (1997). IX—The Opera, 19–20. In the original (1977a: 79): "O grotesco, por exemplo, não está no tetxo do poeta; é uma excrescência para imitar *Mulheres patuscas de Windsor*. Este ponto é contestado pelos satanistas com alguma aparência da razão. Dizem eles que, ao tempo em que o jovem Satanás compôs a grande ópera, nem essa farsa nem Shakespeare eram nascidos. Chegam a afirmar que o poeta inglês não teve outro gênio senão transcrever a letra da ópera, com tal arte e fidelidade, que parece ele próprio o autor da composição; mas, evidentemente, é um plagiário."

32. Caldwell (1960: 165). In what follows, Caldwell quotes Machado's own words: "The French Revolution and *Othello* have been written: still there is nothing to prevent one from lifting this or that scene and using it in other dramas: thus are committed, literarily speaking, acts of plagiarism." *Idem*, 165–6. This passage was extracted from one of the "crônicas" from *A Semana*, published in *Gazeta de Notícias*, July 28, 1895.

33. "Machado julgava necessário que o escritor brasileiro, sem deixar de ser brasileiro, estivesse consciente de que sua obra pertencia a uma tradição universal: a literature". Sá Rego (1989: 5). In this context, it is important to recall José Guilherme Merquior's pioneering essay of 1972, which has been translated into English: Merquior, 1975.

34. John Gledson, "Introduction," in Schwarz (2001: ix).

35. Assis, "To the Reader," (1997: 162). The original (1977b: 260) reads: "(. . .) Conta três fases Humanitas: a *estática* anterior a toda a creação; a *expansiva*, começo de todas as cousas; a *dispersiva*, aparecimento do homem; e contará mais uma a *contractiva*, absorpção do homem e das cousas."

36. In "O alienista" [The Psychiatrist], Machado developed the parody of the scientific discourse to its utmost. There is an English translation (1963).

37. Silviano Santiago. "Latin American Discourse: The Space In Between," in Santiago (2001: 37).

38. I owe this observation to Henning Ritter.

References

Assis, Machado de (1963). *The Psychiatrist and Other Stories*. Trans. William L. Grosmann and Helen Caldwell. Berkeley: University of California Press.

——— (1975). *Quincas Borba*. Rio de Janeiro: Civilização Brasileira/Instituto Nacional do Livro.

——— (1977a). *Dom Casmurro*. Rio de Janeiro: Civilização Brasileira/Instituto Nacional do Livro.

———(1977b) *Memórias póstumas de Brás Cubas*. Rio de Janeiro: Civilização Brasileira / Instituto Nacional do Livro.

———(1997a) *Dom Casmurro*. Trans. John Gledson. New York, Oxford: Oxford University Press.

———(1997). *The Posthumous Memoirs of Brás Cubas*. Trans. Gregory Rabassa. New York and Oxford: Oxford University Press.

———(1998) *Quincas Borba*. Trans. Gregory Rabassa. New York and Oxford: Oxford University Press.

Barth, John (1989). "Foreword to Doubleday Anchor Edition," in *The Floating Opera* and *The End of the Road*. New York: Anchor Books.

Bloom, Harold (2002). "Joaquim Maria Machado de Assis (1839–1908)," in his *Geniuses: A Mosaic of One Hundred Exemplary Creative Minds*. New York: Warner Books.

Borges, Jorge Luis (1989). "Pierre Menard, autor de Quijote." *Obras Completas*, Vol. I. *Ficciones*. Buenos Aires: Emecé.

Brodsky, Joseph (1994). "The Condition We Call Exile or Acorns Aweigh." *On Grief and Reasons*. *Essays*. New York: Farrar, Straus and Giroux.

Caldwell, Helen (1960). *The Brazilian Othello of Machado de Assis*. Berkeley: University of California Press.

———(1970). *Machado de Assis. The Brazilian Master and his Novels*. Berkeley: University of California Press.

Castro Rocha, João Cezar de (2005). The *"Dialectic of Marginality"*: *Preliminary Notes on Brazilian Contemporary Culture*. Centre for Brazilian Studies: University of Oxford, CBS-62-2005.

Cunha, Euclides da (1944). "Preliminary Note." *Rebellion in the Backlands*. Trans. Samuel Putnam. Chicago and London: The University of Chicago Press.

Eakin, Marshall C. (1998). *Brazil: The Once and Future Country*. New York: St. Martin's Griffin.

Fitz, Earl (1989) *Machado de Assis*. Boston: Twayne Publishers.

Fuentes, Carlos (2001). *Machado de la Mancha*. México D. F.: Fondo de Cultura Económica.

Hansen, João Adolfo (1999). "*Dom Casmurro*: Simulacrum and Allegory." in Richard Graham (ed.), *Machado de Assis. Reflections on a Brazilian Master Writer*. Austin: University of Texas Press.

Mac Adam, Alfred (2000). "Review." *Hispanic Review* 68 (Winter): 97.

Merquior, José Guilherme (1972). "Gênero e estilo nas *Memórias póstumas de Brás Cubas*." *Colóquio/Letras*. Lisboa: 12–20.

——— (1975). "A Problematic Vision." *Review of the Center for Inter-American Relations*: 45–51.

Param, Charles (1970). "Jealousy in the Novels of Machado de Assis." *Hispania* 53(2): 198–206.

Rouanet, Sergio Paulo (2005). *Machado de Assis e a subjetividade shandeana*. Centre for Brazilian Studies: University of Oxford, CBS-67-2005.

Sá Rego, Enylton de (1989). *O calundu e a panacéia. Machado de Assis*, a *sátira menipéia e a tradição luciânica*. Rio de Janeiro: Forense Universitária.

——— (1997). "Preface–Warning: Deadly Humor at Work." in Assis, 1997b.

Santiago, Silviano (2001). *The Space In-Between. Essays on Latin American Culture*. Ana Lúcia Gazzola (ed.). Durham and London: Duke University Press.

Sarlo, Beatriz (1988). *Una modernidad periférica: Buenos Aires 1920 y 1930*. Buenos Aires: Nueva Visión.

Schwarz, Roberto (1992). *Misplaced Ideas. Essays on Brazilian Culture*. Edited with an Introduction by John Gledson. London and New York: Verso.

——— (2001). *A Master on the Periphery of Capitalism*. Trans. and with an Introduction by John Gledson. Durham and London: Duke University Press.

Sontag, Susan (2002). *Where the Stress Falls.* New York: Farrar, Strauss and Giroux.

Vilar, Bluma Waddington (2001). *Escrita e leitura: citação e autobiografia em Murilo Mendes e Machado de Assis.* Ph.D. Dissertation, Programa de Pós-graduação em Letras da Universidade do Estado do Rio de Janeiro.

Wagley, Charles (1971). "Preface to Revised Edition." *An Introduction to Brazil.* New York: Columbia University Press.

Zilly, Berthold (2001). "A bárbarie: antítese ou elemento da civilização? Do *Facundo* de Sarmiento a *Os Sertões* de Euclides da Cunha." *Revista Tempo Brasileiro* 144.

Chapter 7

Cuban Cinema: A Long Journey toward the Light

Julio García Espinosa

(Translation and Accompanying Notes by Stephen Hart)

Before anything else I ought to point out that the title of this essay is not intended as a metaphor. It refers to a long journey toward the light of the projector. As you know the first film cameras had a double function, that of filming as well as screening. In Cuba, as in the rest of Latin America, filming and screening have always been separate activities. They have belonged to different teams, different spaces, different interests. To think of them as an inseparable whole has in effect been a long journey toward the light, toward the light of that elusive projector.

The first images created by the new technology arrived in Cuba from France at the dawn of the last century. In 1895 when the Lumière brothers first showed their film reels our country was at war with Spain.[1] The news of their invention, as could be expected, drew the attention of film directors nearby in the United States. So it was that the images of Cuba were among the earliest filmed in the entire world. Since that time we have been trying to maintain a leading role in the creation of our own image.[2]

The first efforts at filmmaking took place in the 1910s when the film industry was still in its infancy. The great precursor was a man who, as one might imagine, had nothing more than his obsession with that magic lantern that seemed capable of defeating death itself. His name was Enrique Díaz Quesada and, despite the rudimentary nature of the tools available to him, he managed to make a considerable number of films. Regrettably, nothing is left of his work since a fire subsequently destroyed everything. All that is left are a few minutes of a documentary that he filmed in an amusement park.[3]

It was in the 1930s and 1940s that Ramón Peón emerged; he managed to make some films following the blueprint that had led to some success in Mexico, Argentina, and Brazil. These successes had been associated with popular music and popular theater, which in those countries as much as in Cuba boasted a long tradition. But Ramón Peón's work did not achieve any widespread support, and his efforts— together with those of many others—were not crowned with success.[4]

In the 1950s, when Italian neo-Realism was at its height, a number of Latin American film directors, including myself, went to study film in Rome.[5] Imbued with the neo-Realist vogue, which favored the creation of an unadorned cinema—that is namely, films without stars or expensive sets—we went back to our respective countries. Just as the long, hard journey toward the light in the world of cinema was taking shape, so too the long, hard journey of the Cold War was beginning.

From the mid-1950s until the fall of the Berlin Wall the Cold War would lead to more than 100,000 deaths in our midst. Latin American film directors suffered persecution, torture; some were murdered and many went into exile.

After finishing our studies in Italy, Gutiérrez Alea[6] and I returned in the mid-1950s to Cuba, when that famous dictator, Batista, was still in power.[7] Even so—and armed with the drop of madness that our predecessors possessed—we decided to attempt to bring cinema to Cuba. And thus it was that, in 1955, the short documentary, *El Megano* (*The Charcoal Worker*), was born. We went to prison as a result. We realized that it would not be possible to make films unless Cuba as a whole changed. And in 1959 our country did change.

With the triumph of the Revolution we tried to resurrect the old dream of bringing together the creation of films and showing them to the public. Producing films required having our own laboratory where we could develop films as well as a sound studio where we could create the sound tape. And it did eventually come about. Film directors, though, would get their training via a process of improvisation. There was not enough time to train them at a film school. We were faced with unrepeatable events and we had to film them, even if the results were not always to a professional standard.

It was in this way that the first documentaries were made. The first two movie features in Cuba were made by Gutiérrez Alea and myself but they were not very successful. They conformed to a rather naive neo-Realism which was, furthermore, outmoded. The great master of neo-Realism, Cesare Zavattini, came to Cuba in the 1960s and we realized that Italian cinema itself had already moved on to another stage.[8] Other film movements—such as Free Cinema, the New Wave,

American Independent Cinema—were emerging, and they were beginning to influence us.[9] During that magnificent decade in the 1960s some of the best film directors in the world came to visit us and we had the opportunity of opening ourselves up to a more plural world. Those were years of enormous innovation. Colonialism seemed to come crashing down to the ground around us. The world was changing and Latin America was as well. The struggle for a true definitive emancipation seemed to be knocking at our doors. It was at the time that what we now know as the New Latin American Cinema came into being. During this period the early Cuban films now regarded as classics were produced: *Memories of Underdevelopment, Lucía, The First Charge of the Machete*, and *The Adventures of Juan Quinquin*.[10] An impressive documentary movement led by Santiago Alvarez also emerged at this time, as did a cartoon tradition pioneered by the film director, Juan Padrón.[11]

Despite all of the aforementioned milestones, achieving the right to show our films in our own country proved to be a traumatic experience. The commercial sector refused to share their jealously guarded freedoms with the film directors. They guaranteed their business profits by opening their doors subject to conditions that were imposed by the big North American companies. They required you to buy ten second-rate films before you were allowed a first-rate film. This prevented the possibility of creating a space for national film production and also prevented us from seeing films from other parts of the world.[12] There was only one way of guaranteeing the existence of a national cinema as well as the equally important right to see foreign films: nationalizing the movie theaters. Flying in the face of all the predictions, movie attendance figures grew sharply during this period. Producing films and screening them—which for more than 50 years had been separate activities—were finally merged. Cuban cinema began to take the long road toward integration within national culture. The past and the present achieved concrete embodiment in the national cinema of this period, and this was brought about by, on the one hand, avoiding the dogma of the apologists as well as, on the other, avoiding the critique of those who were hypercritical of the Revolution.

The long journey toward the light was not exactly a straight line toward the future. At an early stage we had the opportunity of speaking to an African film director. We had spoken to him about our concern that Revolutions always seem to lead to the idea that cinema needs to be based on propaganda. His words of wisdom were as follows: "In Africa we don't have a film industry and, therefore, neither do we have commercial cinema; we don't receive aid from the State and therefore we don't have propaganda cinema. The problem is that we simply don't have any cinema." We understood that just having cinema meant that we would be able to solve the main contradiction in our society. There would inevitably be contradictions but they would involve tears that were less anachronistic as a result.

Of course there were film directors who did not support the socialist agenda offered by the Revolution, and they left Cuba. Those who remained, despite the errors and confusions that arose on the journey, made sure that the independence which we had just won remained a top priority. The elimination of Salvador Allende—although it should be remembered that the latter's political program did not in any case adhere to the Cuban model—alerted us to how important it was to defend our independence.[13] It was for the same cause that film directors in Latin America were fighting. This was why Cuban Cinema became identified with the destiny of Latin American cinema and never adopted the postulates of Socialist Realism.[14]

Indeed the Cuban Film Institute (ICAIC) became the vanguard of Latin American and Caribbean cinema. It is true to say that a number of great Latin American films would not have been made were it not for Cuban Cinema. The creation in 1979 of the Havana Film Festival offered a forum for the annual meeting of film directors. Before that we had only had sporadic—although they were important—meetings in Viña del Mar, Chile; Mérida, Venezuela; Montréal, Canada; as well as an equally important meeting in Pesaro, Italy.[15] But, as a result of the annual meeting at the Havana Film Festival, the Latin American Film Directors' Committee came into being and it effectively consolidated Latin American cinema as a movement providing a broad-based cultural identity for the whole region. Cinema during those years favored the creation of a consciousness of Latin America and the Caribbean as the larger home country (what, in Spanish, we call our "Patria Grande")—something, indeed, which our leaders had proclaimed so frequently.

It was this very Committee that would promote throughout the 1980s the establishment of institutions such as Cinemateques and Film Clubs in Latin America. Finally the New Latin American Cinema Foundation was founded by Gabriel García Márquez, which in turn set up its most important project: the International Film and TV School in San Antonio de los Baños.[16]

The cinematic movement had managed to bring about a situation whereby the film directors were the theoreticians of their own work. It was not a question of defining only one aesthetic model although, in fact, all of us, in our different ways, were attempting to find an aesthetic response to our political positions. It was in this way that the theoretical works of Fernando Birri and Jorge Sanjinés, "Third Cinema" by Fernando Solanas and Octavio Getino, "The Aesthetics of Hunger" by Glauber Rocha, and my own essay "For an Imperfect Cinema" came about.[17]

The disappearance of the Soviet Union and the fall of the Berlin Wall put a stop to this great cinematographic movement whose main aim was to close down once and for all the colonial cycle that we have suffered for more than five hundred years.

Cuban Cinema went during this period from almost twelve full-length movie features a year to no more than four, although even during this difficult stage we continued to produce films as important as *Strawberry and Chocolate* and *Suite Habana*.[18] New generations of Latin American film directors are showing vigorous signs of life; in effect they are embarking on their journey toward the light of the projector. The candle on the horizon has been lit in countries such as Argentina, Brazil, and Venezuela.

What type of films are being made at the moment? We are making any films we can. There are films that are not guided by any particular interest in mind. For what brings us together is the desire to defend the right to show any film—whatever the film is like—in our own countries.

Nevertheless the experiment must go on. We know that our first duty remains that of making Latin America visible. Our countries are invisible countries. A country without an image is a country that does not exist. Death in a country without an image is less painful than a death in a country that has its own image. For this reason, any film, whether experimental or not, is welcome if it makes us more visible.

The search for a more authentic cinema, one that belongs legitimately to the people, even one that is more competitive, must not stop. We do not have stars, nor can we create them. What we need is a cinema in which the character is more important than the star, where people do not leave the movie theater talking about the stars rather than the characters. We need a cinema in which new ways of telling stories do not eliminate the critical spirit of the spectator.

I want to conclude by quoting the last part of my recent brief essay entitled "The End of History." It reads as follows:

> Cinema, as Walter Benjamin would say, loses the aura of the unique and unrepeatable work of art, and thereby de-legitimises the traditional "cult" of the work of art.[19] Looked at from this point of view, cinema

de-sacralizes the relationship with the work and brings about a more open, profane and free communication. This constitutes the essence of its irrevocably popular character.

In the early days cinema, as we all know, disconcerted its audience. It was not clear what to do with it. Was it a spectacle suitable for entertainment? Was it a new visual art? Did the moving image have the status of the fine arts? Some years would go by before it was in effect recognized as the seventh art. Intellectuals and artists from all over the world would rush to legitimize it within the traditional concept of an art form. Europeans would interpret it as an art form. The North Americans saw cinema as an industry. Europeans would rank it according to a predetermined hierarchy as if it were a type of preindustrial art. The North Americans would take advantage of it as if it were a product designed for the masses. Both the European and the North American approach—and even a mixture of the two—would lead to valuable works. Nevertheless both approaches tended to hamper the liberating potential of this new medium. The former would turn cinema into a new cult, but within a traditional framework, while the latter simply degraded it.

It is true, though, that Hollywood from very early on realized that a film could also be a unique and unrepeatable work of art. Hollywood film directors saw that the aura intrinsic to a painting found its equivalent in the unique and unrepeatable charisma of the actor or the actress. In the early days they proved that the masses, as well as a more refined audience, enjoyed the authentic charisma of a Greta Garbo or a Charlie Chaplin.

But Hollywood ruined its own innovation. The profit motive was imposed. As a result a cinema based on actors rather than characters grew up. The star system was born and, along with it, a system of fabricated personalities and immoderate promotions, which tended to create false auras. The spectator went to the movies to pay homage to fame rather than talent. Even great actors and excellent actresses, when they acted out good stories or played complex characters—as a result of the million-dollar promotions—could not escape becoming more important than the characters they portrayed. The industry was the winner. But not only did art itself lose out, the critical spirit of the audience was lost, and the possibility for a freer communication—which is embedded within cinema—was frustrated. These are the rules of the game that underpin storytelling in Hollywood. That is why this "Institutional Representation," to use Noël Burch's phrase, requires more experimentation, another way of telling stories.[20] It needs a cinema whose most striking novelty would consist precisely in transgressing these game rules, thereby opening up the so-called seventh art to the possibility of having a more adult and unalienated spectator. This would be the end of history such as it is narrated nowadays, as well as the end of History with a capital letter, which has mutilated us for too long.

It is neither a question of ignoring the importance of the actor nor of replacing him with special effects, however seductive or fascinating these may be. It is rather a case of seeing acting (and the same might be said of photography, music, and *mise-en-scène*) as simply one element among many enhancing the significance of the characters and consequently the plot.

These ideas are just as applicable to filmmaking which, like ours, cannot rely on—nor has it any realistic hope for economic reasons of creating—a star system. Using the same paradigm as that which underpins the star system without actually having a star system is not only inappropriate; it implies that we are still lost in the labyrinth of incompetence and controlled markets.[21]

The identity of cinema is in crisis, and film festivals should reflect this situation. Recognition should be given to the "most un-alienating film" and to the "best character" rather than the "best actor," even to the "worst-illustrated novel." It is fundamentally also a question of believing that art is not something to compete over. Art is for sharing with others.

The problem is that, in its own way, cinema has brought about a crisis within art as a whole. Nowadays the new technologies are subverting more than ever the traditional concept of art. Cinema can still be called the seventh art and its history can be written in the same way that the history of the fine arts can be written. Nobody thinks of television as the eighth art, nor of telling its history with the aid of traditional paradigms. Cinema has its own form of museum in the cinematheques. Yet TV does not have its own "telematheques." And even more disturbing is the appearance of the computer or the PC. It is not by chance that there is so much talk nowadays of the demise, or the crisis, of art. But what is art today? It was easy to say when these new technologies did not yet exist. It will be a vain project to attempt to hierarchize these new technologies, paying them a homage that is, indeed, alien to them. The transition from the sacred to the profane seems irreversible.

Notes

1. After the "Ten Years War" (1868–78) in which Cuban nationalists failed to rally support for their cause, a new insurgency broke out in 1895 spearheaded by José Martí, calling for Cuba's liberation from

Spain. The Lumière brothers (Auguste and Louis) are credited with the world's first public screening of a film; they demonstrated their cinematograph on December 28, 1895 at the Grand Café in Paris, and it quickly spread to the rest of the world.

2. For further discussion of the evolution of Cuban film see Chanan (2003) and García Osuna (2003).

3. Enrique Díaz Quesada created the first Cuban movie film, *Un duelo a orillas del Almendares* (A Duel on the Banks of the Almendares; 1907), as well as a documentary, entitled *El epílogo del Maine* (The Epilogue of the Maine; 1912), about the aftermath of the mysterious explosion of the *USS Maine*, which had occurred in Havana harbor in April 1898, sparking the Spanish-American war and eventually causing Spain to lose its last major overseas colonies: Cuba, Puerto Rico, and the Phillipines. Neither of these reels have survived; the only footage by Díaz Quesada that has survived is a one-minute documentary, *Parque de Paletino* (Paletino Park); see García Espinosa (2002: 261–2).

4. Ramón Peón (1897–1971), known by some as the "Cuban Griffith," was unable to make films in his native Cuba and went to Hollywood and then Mexico to direct films, the most notable of which was *Romance del Palmar* (Romance in the Palm Grove, 1938), starring Rita Montaner.

5. This was at the Centro Sperimentale di Cinematografia in Rome, which was attended not only by García Espinosa and Gutiérrez Alea from Cuba but also Fernando Birri (Argentina) and Gabriel García Márquez (Colombia).

6. Tomás Gutiérrez Alea (1928–96) was one of the key figures of Cuban cinema in the post-revolutionary era. He was cofounder—along with Julio García Espinosa—of the Cuban Film Institute (ICAIC), directing such classics as *Memories of Underdevelopment* (1968), *The Last Supper* (1977), and *Strawberry and Chocolate* (1993).

7. Fulgencio Batista was ousted from power by Fidel Castro; Batista fled Havana on December 31, 1958, and Castro took over the following day.

8. Cesare Zavattini (1902–89) was one of the major Italian film directors associated with neo-Realism. For a discussion of neo-Realism, see Marcus (1987).

9. Free Cinema was a groundbreaking documentary movement that emerged in Britain in the 1950s; the term was first coined by Lindsay Anderson in 1956. The New Wave, or Nouvelle Vague, was a French film movement of the late 1950s associated with directors such as Louis Malle, Claude Chabrol, François Truffaut, Alain Resnais, and Jean-Luc Godard. For further discussion, see Marie et al. (2002). For more information on American Independent Cinema—a film movement that blossomed in the United States in the 1970s and 1980s and which was a reaction against the Hollywood studio system—see Hillier (2001).

10. *Lucía* (1968) was directed by Humberto Solás, *La primera carga al machete* (1969) by Manuel Octavio Gómez, and *Las aventuras de Juan Quinquin* (1963) by Julio García Espinosa.

11. Santiago Alvarez (1919–98) was renowned for his short documentaries, which often focused on pressing social issues. Juan Padrón Blanco (b. 1947), cartoonist and film scriptwriter, is perhaps best known as the creator of the cartoon character, Elpidio Valdés.

12. For further information on Hollywood's marketing and distribution strategies during this era, see Schnitman (1984).

13. Salvador Allende came to power in Chile in 1970, the first ever democratically elected Marxist president of a Latin American country, but he was removed from power by the *coup d'état* led by Augusto Pinochet, the Head of the Armed Forces, three years later, leading to a right-wing dictatorship that ruthlessly repressed the working class.

14. Socialist Realism was a politically committed art associated with the Soviet Union and its satellites. It was characterized by the expression of a clear, left-wing political message to which artistry and style were deemed to be either secondary or redundant. For further discussion see Lahusen (2002).

15. The first meeting of the Latin American Film Festival was held in Viña del Mar, Chile, in 1967, then in Mérida, Venezuela, in 1968, then back in Viña del Mar in 1969 (see Hart 2004: 8–9). The Pesaro Film Festival was founded in 1965, the Montréal World Film Festival in 1977, both of which were responsive to the New Latin American Cinema, but it was not until 1979 that the Latin American Film Festival had a permanent home in Havana, sponsored by the Cuban Film Institute.

16. García Márquez established the Fundación de Nuevo Cine Latinoamericano in 1985, which led to the creation of the Escuela Internacional de Cine y Televisión in San Antonio de los Baños which Julio García Espinosa currently directs.

17. These essays are now available in English in Chanan 1983.

18. *Fresa y chocolate* (1993) was directed by Tomás Gutiérrez Alea and *Suite Habana* (2003) by Fernando Pérez.

19. See Benjamin's essay, "The Work of Art in the Age of Mechanical Reproduction" (1970).

20. See Noël Burch's *Life to those Shadows* (1990).

21. For further discussion of the star system in Latin American film, see King (2003).

References

Benjamin, Walter (1970). "The Work of Art in the Age of Mechanical Reproduction," in his *Illuminations*. New York: Vintage, pp. 211–35.

Burch, Noël (1990). *Life to Those Shadows*. Berkeley: University of California Press.

Chanan, Michael (ed.) (1983). *Twenty-Five Years of the New Latin American Cinema*. London: BFI, in conjunction with Channel 4.

——(2003). *Cuban Cinema*. Minneapolis: University of Minnesota Press.

García Espinosa, Julio (2002). *El cine cubano: un largo camino hacia la luz*. Havana: Casa de las Américas.

García Osuna, J. (2003). *The Cuban Filmography. 1897 through 2001*. New York: McFarland and Company.

Hart, Stephen M. (2004). *A Companion to Latin American Film*. London: Tamesis.

Hillier, Jim (ed.) (2001). *American Independent Cinema: A Sight and Sound Reader*. London: BFI.

King, John (2003). "Stars Mapping the Firmament," in Stephen Hart and Richard Young (eds.), *Contemporary Latin American Cultural Studies*. London: Arnold, pp. 140–50.

Lahusen, Thomas (2002). *How Life Writes the Book: Real Socialism and Socialist Realism in Stalin's Russia*. Ithaca: Cornell University Press.

Marcus, Millicent (1987). *Italian Film in the Light of Neorealism*. Princeton: Princeton University Press.

Marie, Michel, Richard John Neupert, and Richard Neupert (eds.) (2002). *The French New Wave: An Artistic School*. Oxford: Blackwell.

Schnitman, Jorge A. (1984). *Film Industries in Latin America: Dependency and Development*. Norwood, NJ: ABLEX.

Chapter 8

Culture and Communication in Inter-American Relations: The Current State of an Asymmetric Debate

Néstor García Canclini

(Translated by Stephen Hart)

When, how, and where was, or is, Latin America modern? Answering this question, as we know, implies entering into that debate about what we understand by terms such as "modernity," "modernization," and "modernism." The task of understanding Latin America was understood throughout the nineteenth century as a search to understand the contradictions between, on the one hand, an exuberant cultural modernism and, on the other, a deficient modernization. It was also a question of deciphering how it was that modernization, which had been accelerated by the twin processes of industrialization and urbanization, existed side by side with archaic traditions. The different paradigms of modernity with which these contradictions were analyzed almost always had one thing in common: they were conceived within a national context. The fundamental question was as follows: How are Brazilians, Peruvians, and Mexicans able to live in modern nations, and what can they do with those throwbacks (*rezagos*) or hybridizations that persist in exhibiting non-modern features?

In recent years the space of the nation has become blurred; it is no longer the backdrop against which modernization occurs. To be modern, nowadays, is to travel, communicate, exchange with the world. Goods, messages, and people are considered to be modern if they circulate globally, if they speak various languages, and are attractive in a high number of markets.

For that reason I shall be investigating the ways in which we Latin Americans are modern in relation to circulation and globality. Perhaps the specific place we inhabit, like that of other peripheral regions, makes clear that a key characteristic of the current stage of globalized modernity is that goods and messages travel with greater ease than people do. To test this hypothesis I shall focus on migration patterns and intercultural communication via various media.

Migration at Different Stages of Modernity

For Latin America, modernization was associated with the international circulation of people and communications. But not in the direction that is familiar to us today. Emigration from Europe to the American continent was, indeed, a foundational element of our modernity. There was a significant period from 1846 until 1930 when some 52,000,000 people left Europe. Twenty-one percent of those emigrants traveled to Latin America: there were approximately 10,000,000 people, 38 percent of whom were Italian, 28 percent Spanish, and 11 percent Portuguese. The majority of these Latin emigrants chose Argentina as their favored destination, followed by Brazil, Cuba, the Antilles, Uruguay, and Mexico. If we bear in mind that at the beginning of the twentieth century the total population of Europe was some 200,000,000 people, this means one quarter of the population left. The arrival of these emigrants in America during the period from 1840 until 1940 led to an increase in Argentina's population of the order of 40 percent; the percentages for population growth at this time are 30 percent for the United States and approximately 15 percent for Canada and Brazil (González Martínez 1996). It is well known how much this influx from Europe contributed to the modernization of industry, the development of the educational system, the creation of publishing houses, in short, the designing and implementation of modern nation-building projects.

What has been happening in the last few decades? Migration patterns are different nowadays. Migration in the nineteenth century and the first half of the twentieth century was almost always permanent and led to the cutting of ties between those who left and those who stayed behind, whereas the notion of population movement nowadays encompasses permanent and temporary relocation, as well as short journeys for the purposes of tourism or work-related activities.

Three types of migration can be distinguished nowadays: (1) migration for the purpose of permanent settlement or population; (2) temporary migration for work reasons; and (3) migration that involves a relocation of variable status, and which is midway between the two previous types of migration. These latter two are the types of migration that have increased in recent decades (Garson and Thoreau 1999). The ebb and flow of migration is controlled and subject to restricted duration and restricted conditions. Unlike permanent migration, which was linked in the past to the policy of population, in recent times many residence permits are temporary and discriminate on the basis of nationality and the economic needs of the host nation. Authorization to remain in the country may be renewed but those countries that are the most attractive and have the greatest amount of migrants (normally identified as the industrialized Western countries) only grant nationality to a small minority and, furthermore, limit the rights, stability, and integration of foreigners in the host country. Even when the migrants are accepted because their work expertise coincides with the needs of the economy adopting them, sociocultural short-circuits still occur: segregation within certain districts, denial of access to schools and health services, as well as the negative evaluation of certain beliefs and customs, which can lead to aggression and even deportation.

These trends vary between countries—which have different policies—and also vary according to the classification of migrants: professionals, technicians, intellectuals, and specialized workers are traditionally more welcome. It is rare for the right to travel of the rich and the well educated to be questioned. Those who have a fat checkbook, arms smugglers or drug traffickers, as well as the bankers who launder their money for them, as Hans Magnus Enzensberger suggests, "do not have prejudices" and "are above nationalism" (Enzensberger 1992: 42). Nevertheless, the instability that is common to all labor markets as a result of globalized competition highlights the uncertainly underlying the status of foreigners and makes their integration into the host society difficult (Garson and Thoreau 1999).

As a counterweight to these disadvantages for migrants nowadays the possibility of keeping a fluid communication with their country of origin has been enhanced. Daily newspapers from Europe arrive in the capital cities of Latin America while free-to-air and cable TV allows access to channels from Europe and the United States. Audiovisual media, email, family networks, and friends networks have changed contact between the continents from what used to take weeks or months in the past into a constant activity nowadays. Disembarking is not the same as landing, nor physical travel the same as electronic navigation. Interculturality is created nowadays more as a result of communication via email rather than through the physical relocation of the migrant.

In order to see with greater clarity how the phenomenon of migration has changed it is also important to recall that in the second half of the twentieth century the direction taken by the migrant has been reversed. Between 1960 and 1965, Argentina, Venezuela, Brazil, and Uruguay received 105,783 Spanish emigrants. But in the following two decades more than 1,000,000 Spaniards preferred to emigrate to other European countries (González Martínez 1996). At the same time a new cycle of emigration from Latin America to Spain, Italy, and Germany began, as well as to a lesser extent to other European countries. These émigrés were made up of millions of individuals who were politically persecuted, or unemployed, or people who were tired of the limited horizons offered by countries in the Southern Cone or Central America. The period in which Europeans could "make it rich in America" (*hacer la América*) had effectively come to an end and it ushered in a new era in which South Americans (the so-called *sudacas*) were willing to contemplate becoming part of Europe's economic growth.

It is possible to hypothesize that the exchanges which occurred in the nineteenth and twentieth centuries should have modified the polarity created between Europe and America during the Conquest and the Colonial period. Nevertheless certain stereotypes can be observed to have persisted: the discrimination of Europeans toward Latin Americans, the admiration and distrust of Latin Americans toward Europeans. The transformation of the links, in effect, simply reproduced a long-lasting asymmetric structure. This is evident in the limits placed on entry or, alternatively, the ease with which entry is obtained by others.

Why have laws become so restrictive for Latin Americans in European nations as well as the United States? When human rights movements question these restrictions the response is that migrants can no longer be accepted in the same way as occurred when the Americas had immense territories to populate and when they saw the new arrivals as an incentive to develop industry, education, and modern services. Furthermore we are told that in Europe and the United States where there are already millions of foreigners, unemployment has grown in recent years. Many sectors of society have indeed gone as far as to blame migrants for the increase in delinquency and social conflicts (Dewitte 1999).

Even though many things have changed from the nineteenth to the twentieth century, a decisive change in this process of interaction has been that capital, goods, and emails pass from one country to another more easily than people do. It is easier to invest in a foreign country than it is to become a citizen of that country. The free-trade agreements, which are promoted as the engine behind modernization, almost never include the notion of the universalization of human rights—which is intrinsic to modernity—including the rights of those people who are different as a result of being migrants. We have moved from an enlightened modernity to a neoliberal modernity.

Another radical change in recent decades has been the substitution of Europe by the United States as the referent for modernity. Latin America—which up to a point was a European invention—now finds its otherness mainly in U.S. society and the U.S. empire. The figures are well known: some countries—like Mexico, for example—have 90 percent of their trade with the United States. Several Latin American nations lost 10–15 percent of their total population to the United States, such that nowadays Spanish speakers number more than 40,000,000 in the United States.

The money sent home by migrants living in the United States—from Mexico, the Dominican Republic, and El Salvador—became the principal net source of hard currency for their respective countries of origin. In 2004 the money sent back to Mexico by Mexican emigrants reached $16,613,000. The currency sent home has had more of a significant impact on the rural and urban economies of Mexico, the Dominican Republic, and El Salvador, and their families living there—and their experience of modernity—than any of its exports.

Imaginaries and Intercultural Misunderstandings

How can we reconceive the process of modernization within the current phase of globalization? It is well known that these changes of perspective are the result of socioeconomic transformations, as revealed by the facts and figures to which I have just alluded, and are also the result of the new imaginaries that guide the social actors. Before showing some documentary and artistic images that allow us to visualize this process, I should like to propose a rereading of a classic image of Latin Americanness: the map of South America, which Joaquín Torres García drew in 1936.

"Our north is the south," Torres García declared in his manifesto. Putting the map upside down encouraged us to conceive of the world from our own nation or city, from Montevideo, for example. That inverted map could be read nowadays as a metaphor of a Latin America that points the needle of its compass toward the north where it imagines life to be better; for migrants that better life is not to be found within one's own nation. Alternatively we could interpret

Torres García's metaphor as representative of the asymmetric bidirectionality of cultural exchange.

I want to take up, in this sense, the opposition between two recent artistic works which I analyzed in my recent book, *La globalización imaginada* (1999). The first is a work entitled *América*, by Yukinori Yanagi, consisting of 36 flags from different countries, made out of small plastic boxes full of colored sand. The flags are joined together by tubes along which ants travel, thereby wearing away and mixing up the flags. Yukinori Yanagi created the first version of this work in 1993 for the Biennial Exhibition in Venice. In 1994 he made a replica in San Diego, in the context of a multinational art exhibition called inSITE, made up of flags from the 3 Americas. After a few weeks the emblems became unrecognizable. Yanagi's work can be interpreted as a metaphor of those migrant workers who are gradually deconstructing nationalisms and imperialisms all over the world. But not everyone who saw the exhibit noticed this. When Yanagi presented this work at the Biennial Exhibition in Venice, the Animal Protection Society managed to close it down for a few days, stopping the artist from conducting his "exploitation of ants." Other reactions were the consequence of the fact that the public did not like seeing the differences between nations destabilized. Yanagi, for his part, was attempting to express his experience of the point at which the marks of identity dissolve. The species of ant, which was obtained from Brazil for the Biennial in São Paolo in 1996, seemed too slow for him and, when the exhibition opened, he expressed his fear that the flags would not be sufficiently transformed as a result.

This metaphor suggests that migration on a massive scale along with globalization should change today's world into a systems of flows and interactivities in which the differences between nations would eventually be dissolved. Demographic data, however, do not bear out this image of total fluidity, nor even that of a pervasive transnational mobility. The total number of people who leave their countries in order to settle in another country for more than a year varies between 130,000,000 and 150,000,000, which is on average 2.3 percent of the world's population. "Our 'nomadic planet,' in which people move around more and more rapidly—as Gilda Simon points out—while it costs less and less to do so, is in point of fact full of sedentary people; the image of a world covered by uncontrollable waves of migration belongs rather to the grand shop of clichés" (Simon 1999: 43).

There is another way of understanding the exchanges between the United States and Latin America in that emblematic city of Tijuana, the most frequently crossed border in the world. More than

90,000,000 people cross between Tijuana and San Diego every year to enter or leave the United States. Many are migrants and others are workers who live in one city and work in the other. Furthermore, more than 40,000 tourists visit Tijuana every day and 45 percent of them remain for less than 3 hours in the city; in the last 14 years the population of the city has doubled. How does an artist from Tijuana represent this bidirectional exchange? We see it in the Trojan horse erected by Marcos Ramírez Erre in the latest edition of the inSITE urban art program created in 1997, between Tijuana and San Diego. The artist erected, a few meters from the stalls of the border, a wooden horse that was 25 meters high, with two heads, one looking toward the United States, the other looking toward Mexico. In this way it avoids the stereotype of one-directional penetration going from north to south. It also avoids the opposite illusion of those who state that migration from the south is smuggling something into the United States without their realizing what is going on. The artist told me that this fragile and ephemeral "anti-monument" is "transparent because we already know what their intentions are with regard to us, and they know what our intentions are with regard to them." Amid the Mexican vendors who wander between the cars that are piled up in front of the stalls which used to offer Aztec calendars or Mexican handicraft and now are simply an addition to "Spider Man and Walt Disney toys," Ramírez Erre did not present a work with a nationalist ethos, but rather a modified, universal symbol. The alteration of the Trojan horse as a commonplace of historical iconography led to its transformation into a symbol indicating the multidirectionality of messages as well as the ambiguities which their use in various media can lead to. The artist reproduced the image of the horse on T-shirts and post cards so that it could be sold alongside the Aztec calendars and the "Walt Disney toys." He also had four Trojan costumes so that anyone who wanted to have himself photographed next to the "monument" could do so, thus creating an ironic allusion to the photographic images that tourists routinely create next to symbols of Mexicanhood and the American way of life.

Yanagi's ants that deconstruct the flags suggest a pervasive interaction whereby the very marks of identity would eventually disappear. As far as the two-headed horse is concerned, it represents the bidirectionality and reciprocity of interactivity; the transparent character of the animal suggests that "what they want from us and what we want from them" can no longer be hidden; the conflict has become explicit, but it is not depicted via nationalist imagery but rather with a multinational symbol, which, when reread, invites us to reflect about a specific border. While Yanagi's work celebrated the dissolution of national barriers, Ramírez's two-headed horse and its ensemble of performance installation (T-shirts and Trojan costumes to put on and take a picture of yourself in, souvenirs that parody the neo-handicraft designed for tourist consumption), situated as it is on the actual border between the United States and Mexico, demonstrated how intercultural misunderstandings are created.

Contradictions in Multiculturality in the North and the South

Finally I want to focus on what the cultural industries tell us about inter-American relations. In general terms the mass media manage to circulate their messages with greater ease than is permitted to individuals. While migration, and the sending of money and narratives back home allow us to interact with one another as well as to receive more information from other countries than was possible in any previous era, their communicational power cannot be compared to that offered by radio, television, cinema, and the Internet. Intercommunication in these fields demonstrates the asymmetry between north and south, as well as the unequal nature of the opportunities available to participate in that globalized modernity.

How do we see North Americans in Latin America and how do they see us? I shall take cinema as a test case. In the 1960s, 10 percent of the films circulating in the U.S. market were imported. Nowadays the figures have dropped to 0.75 percent. The meager diversity of what is offered on screen is due to various factors: the corporate organizations behind screening in theaters; the increase in real estate costs and promotion costs for distributors and exhibitors; the pervasive self-satisfaction of North Americans with regard to their society, language, and lifestyle, as well as the resistance by the masses to the idea of relating to other cultures or their goods.

The contradictions underlying this almost monolingual policy in the media are plain to see, even when discounting a few exceptions made for other languages—as a result of the multilingual and multicultural character of U.S. society. The last census listed the United States as having 35,000,000 Spanish speakers, namely 12 percent of its total population—and 63 percent of these are of Mexican origin. The percentage of Spanish speakers is even higher in cities such as Los Angeles (6,900,000) and New York (3,800,000). Miami, Chicago, Houston, and the San Francisco Bay area all have around 1,500,000 Spanish speakers each. For this reason it does not take much imagination to see how receptive these population groups would be to Spanish language films or films made in Latin America (Miller 2002).

The predominance of U.S. films within the United States—which almost completely excludes other filmic traditions—is echoed, in a startling way, in Latin American countries. Even in countries that have a long tradition of national filmmaking, such as Argentina, Brazil, and Mexico, Hollywood movies take up around 90 percent of screen time. In many European countries and on other continents, as we already know, the situation is a similar one.

The global hegemony of U.S. cinema "came about historically as a result of clearly political factors," although these factors in principle and—judging by appearances—were fortuitous; factors such as the two world wars, which destroyed the filmmakers who were in competition with them, along "with the active support of the U.S. government." "The global predominance of the United States in the cultural and audiovisual industries does not have one cause, just as it was not of course the result of 'spontaneous combustion.' It was an historical result caused by a number of factors" (Sánchez Ruiz 2002: 23). At the same time it is important to add that the new benefits provided to foreign investment as a result of the deregulation policies adopted by Latin American governments from the 1980s onward also played their role; they led, for example, to sustained U.S., Canadian and Australian investment in the construction of multiscreen theater complexes in large and medium-size cities throughout Latin America. Transnational capital in this way controlled screening, making it uniform and favoring internationally successful films, which thereby reducing screen time for other filmic traditions. Comparative studies of films screened in Latin American capitals demonstrate that in the last 40 years screen space has increased but the variety of films offered has diminished. In Mexico in 1990, 50 percent of the films screened were U.S. and 45.6 percent were Mexican. By 2000, the ratio had changed to 84.2 percent U.S. as against 8.3 percent Mexican. In 1995, the year when the expansion of multiscreen theaters began, 16.8 percent of the film screened were neither American nor Mexican; by 2000 this figure had dropped down to 7.5 percent (Rosas Mantecón 2002).

Other factors have contributed to this predominance of U.S. cinema: (a) the early development of the film industry in the United States (which was parallel to developments in the fields of culture and communications), which generated an accumulation of professional experience, sophisticated technical knowledge, and an advanced knowledge of the markets; (b) rapid urbanization and industrial development, in the United States and Latin America, which led to strong migratory patterns; (c) tax exemptions as well as other protectionist incentives used by the U.S. government to aid its national film industry, combined with a semi-monopolizing control over distribution and screening, which itself became more effective as a barrier against the film industries of other countries and other languages than the screen quotas that were established in other countries via the regulation of public organizations (McAnany and Wilkinson 1996).

We see once more in cinema this divergence between, on the one hand, ways of conceiving social multiculturalism within the United States and, on the other, a policy of rejection of diversity in the cultural industries, which operates as much within the space of the nation as in the control of international markets. The United States is the country that has most forcefully backed "Affirmative Action," that is, the granting of privileged conditions for minorities who are excluded or marginalized within the nation. At the same time, however, the United States pursues an aggressive policy of marginalization of the diversity of goods and cultural messages that come from outside its territory, via transnational circuits such as cinema, television, and music—which are, indeed, managed by U.S. companies. This marginalization also occurs in the context of international organizations (WTO, UNESCO, etc.), where the United States opposes any action that protects the cultural industries of other nations. This one-dimensional approach is also evident in the undervaluing of expression by the minorities—whether in art or in the media within the United States.

Many artists have expressed in their work—whether their art is visual, plastic, or literary—this sense of the unequal interaction between the United States and Latin America, as well as the sociocultural consequences of this inequality for inhabitants on each side of the fence. I wish to pick up here on the photograph and text by Allan Sekula about the filming of the film, *Titanic*, in the Mexican sea, near Rosarito, which Sekula presented in his exhibition for inSITE in 1997. The sinking of the Titanic was filmed by Universal Studies in Popotla, a beach to the south of Tijuana, in order to take advantage of low wages in Mexico (they are ten times less than in the United States). Sekula sees this "intervention" as part of a continuous process of actions going back to 1840 by "white adventurers" who came to Baja California, "an inferior space, a utopia of child-like freedoms where the lobsters can be eaten up greedily and where cars can be driven with careless abandon. And now, Hollywood itself is fleeing, it is crossing the triple-layered fence in order to expose its own and very dear vision of the history of a modernity which stumbles upon the primordial abyss." He continues: "The extras float and shiver among the dead body dummies, gesticulating and choking according to orders, a real army of people drowning . . . the industrial border to the north of Mexico is the prototype of a dark Taylorist future."

The *Titanic*, Sekula suggests, "is the old precursor of an unknown machine-operator (*maquiladora*). An army of cheap labourers is contained within and directed by the hydraulic action of the machinery of apartheid. The machine is becoming more and more indifferent to democracy, on both sides of the line, but it is not indifferent to culture, an oil sprinkled on murky waters."

Combinable Options

The analysis of migratory patterns, as well as the asymmetry between north and south in cultural exchanges and communication exchanges, demonstrates that the redefinition of the modern is operating in a globalized and unequal way. The axes of the question are not so much being articulated around the modernization and traditions within each nation; rather they are taking shape in the ways in which this and that region—with their distinct ways of being modern—are repositioning themselves in the context of global exchanges.

Three years ago, when I published a more detailed exposé of these matters in my book, *Latinoamericanos buscando lugar en este siglo* (Paidós), the cover designer, Mario Eskenazi, created a map of Latin America which was decentered and multicolored. He did not invert the south and the north, as in Torres García's concept; he interpreted the mobility that is occurring in our continent nowadays as a blurring of the borders and a superimposition of planes. Our map is now black, red, yellow, green, blue, and mauve; it multiplies itself and spreads itself through space. The identities within Latin America are multichromatic and are not fixed in one place. Latin America is not contained simply within the territory which we are accustomed to designate with that name. There are millions of Latin Americans in California, New York, Madrid, London, or Paris, and our cultural products—novels, soap operas, scientific studies, and music—are searching for their place in every continent. It is a modernity that is decentered or eccentric.

Perhaps the discrepancy between these various images, these various ways of imagining Latin America, corresponds to the various alternatives that suggest how we should face the future. Personally I believe that it is not a question of a dilemma understood in absolute terms, but rather a question of how we should combine two necessary tasks: Torres García's proposal to place the north where the south is via a map that explores this eccentric and multicolored place. There are many ways of being Latin American in our world.

Nevertheless it also must be said that if we place ourselves within the simple legitimacy offered by differences, if we only recognize the many ways in which one can be Latin American (as an Indian, an Afro-American, a white, etc.), then we are not facing up to the growing inequality created by asymmetry. Multiculturalism—whether canonized in the menu offered by many museums, publishing houses, music companies, or TV companies—is administered via a funnel system whose seat of power is located in a few centers in the north. The new strategies for dividing up artistic and intellectual work, the accumulation of symbolic and economic capital via culture and communication, lead to a situation whereby the wealth of almost the whole planet—along with the ability to capture and redistribute diversity—is concentrated in the United States, some European countries, and Japan.

The global expansion of economic and cultural exchanges, migration that has spread in all directions, and informational links across the globe, work against the respectful relativism that occurs in the context of specific, isolated cultures. When the borders between groups, ethnicities, and nations become so blurred and unstable, and when competitiveness leads to anger, at that point, a humanist tolerance—as a simple ethical term—is inadequate.

We are beginning to find out what a globalized citizenry would be like. At this level of effective participation by citizens, the issue arises of the decisive importance of politics as action carried out by society, not simply in terms of agreements between high-ranking officials or participation simulated via the media. Neither should it be the mere resistance of actors or disparate movements. In a world that is designed at once to interconnect as well as to exclude, the two most tried and tested policies to date for interculturality—tolerance toward people who are different and solidarity with the subalterns—are both necessary in order to allow us to carry on living together with one another. But if they stop there they run the risk of becoming resources allowing us to live with what we are not allowed to do. In Latin America, as elsewhere, communicating with people who are different, fighting inequality, and making sure that access to intercultural heritage is available to all, have become indispensable tasks so that we can finally escape this era of paltry abundance.

References

Dewitte, Philippe (1999). *Immigration et intégration: l'état des savoirs.* Paris: La Découverte.

Enzensberger, Hans Magnus (1992). *La gran migración.* Barcelona: Anagrama.

Garson, Jean Pierre and Cécile Thoreau (1999). "Typologie des migrations et analyse de l'intégration." In Dewitte.

García Canclini, Néstor (1999). *La globalización imaginada.* México: Paidós.

—— (2002). *Latinoamericanos buscando lugar en este siglo.* Buenos Aires: Paidós.

García Canclini, Néstor, Ana Rosas Mantecón, and Enrique Sánchez Ruiz (eds.) (2002). *Cine mexicano y latinoamericano: situación actual y perspectivas en América Latina, España y Estados Unidos. Informe presentado al Instituto Mexicano de Cinematografía,* diciembre.

González Martínez, Elda E. (1996). "Españoles en América e iberoamericanos en España: cara y cruz de un fenómeno," *Arbor* 154: 607.

McAnany, E. G. and Wilkinson, K. T. (eds.) (1996). *NAFTA and the Cultural Industries.* Austin: University of Texas Press.

Miller, Toby (2002). "El cine mexicano en los Estados Unidos." In García Canclini, et al.

Rosas Mantecón, Ana (2002). "Las batallas por la diversidad: exhibición y públicos de cine en México." In García Canclini et al.

Sánchez Ruiz, Enrique (2002). "La industria audiovisual en América del Norte: entre el mercado (oligopólico) y las politicas públicas." In García Canclini et al.

Simon, Gilda (1999). "Les mouvements de population aujourd'hui." In Dewitte.

Conclusion

When Was Latin America Modern?

Laurence Whitehead

This volume arises from a conference, and a collaborative investigation, under the challenging title "When was Latin America modern?" So when was it? Was it modern when, ca. 1500, Europeans equipped with the latest technology in navigation, for the first time, not only succeeded in crossing the Atlantic from East to West, but also in retracing their steps and plotting out the routes their successors could reliably follow to repeat their pioneering achievements? Or was it modern when the population of Haiti launched into the first success-ful slave revolt since Antiquity, in 1793? Or perhaps it was when, for the first time, digital satellite mapping generated a uniform and com-prehensive picture of the state of the Amazonian forest (and its rate of deforestation)? Or was it when the latest developments of the princi-ples of impersonal rule-bound self-government ("republican constitu-tionalism") were enshrined in the foundation documents of the "modern-style" nation-states created so precociously out of the frag-ments of the Iberian mercantile empires, at the beginning of the nine-teenth century? Did the coming of railways signal the advent of modernity; or was it heralded by the internal combustion engine, the contraceptive pill, or the Internet? Was the reconstruction (or out-right invention) of a nationalist historical tradition to be propagated throughout the jurisdiction by state-appointed and regulated teachers the acme of modernity? Or was it the enshrinement of a complex of "traditional" blinkers and constraints that then became a barrier against the more truly modern "advanced" societies of the Northern hemisphere? Or perhaps all this is too Eurocentric, and the authenti-cally "modern" moments in the history of the subcontinent were signaled by other achievements—the domestication of maize, or the astronomical discoveries embodied in the Mayan calendar.

All of these were "moments" of modernity, forward-looking developments, and discoveries in the context of their times. Reference to a "moment" implies both the force of this progressive impulse, and its impermanence. My view, which comes under the generic label of "multiple modernities," contrasts with the triumphalism and implied teleology of 1960s modernization theory. It assumes that there is no single necessary path leading to a preordained and presumptively superior "modern" outcome. And it takes it as given that modernity will take a variety of contested forms, none of them definitive and that the drive for modernization carries heavy and unequally distributed costs. Modernity cannot, for example, be reduced to the geographical and temporal specifics that some historians have attributed to the so-called Enlightenment Project,[1] *both* because it transcends that period *and* because both of this pair of terms condense multiple possibilities. Indeed, depending on the context, the process of modernity can destroy more value than it creates. The task of this chapter is to flesh out that approximate idea, as it applies to the large region of "Latin American and the Caribbean"[2] over the *longue dureé*. Of course, it can be no more than a synoptic overview, an exploratory sketch drawing on insights of the various contributors who participated in this joint investigation. What follows is an exploration and synthesis of their substantive contributions, and a reflection on what they suggest about Latin America's distinctiveness and about "modernity" as an object of enquiry.

On a purist definition of what it is to be "modern" the obvious answer to the question "When was Latin America modern?" would be "never." But from a multiple modernities perspective it would be more persuasive to answer "always"—at least since independence. That is the case made in a recent book[3] and can be sustained by the various contributions to this volume. It is evidently quite impossible to overlook the hybrid and heterogeneous nature of all aspects of the social, political, economic, and cultural realities of the subcontinent. Even the most widely diffused of "modern" characteristics (the audience for telenovelas, or the propensity for international travel and communications, for instance) only achieve partial coverage. Those engaged in such activities coexist side by side with other important segments of the population whose perspectives remain resolutely face-to-face, parochial, immobile, and indeed, to use the conventional polarity, "traditional." The abstract conception of a totally "modern" society is, therefore, flatly inapplicable to our region, even if it could be held to approximate reality in some other part of the world. But where would that be? Singapore? Silicon Valley? The City of London?

All the obvious candidates for a strict embodiment of modernity seem quite problematic, not only because they too contain traces of traditionalism, but also because they are such limited and partial components of a more hybrid totality. So the answer is never easy to understand and requires elaboration. What requires more careful discussion is the alternative "multiple modernities" perspective, according to which Latin American and Caribbean can be said to display a broad and recurrent "bias towards modernity" rather than undergoing linear and cumulative progress from premodernity to modernity, and even postmodernity. This large region has always "been modern"—it has always tended toward modernization and modernity according to this less purist conception. This claim supports the alternative answer to the opening question.

So what is this "approximate idea" of modernity, and why is it worth defending and using to understand the distinctive characteristics of the subcontinent? One essential component is the understanding that one's own community or society, however familiar and supportive it may be, is only a small component of a much larger international or indeed global system, so that even when the local is traditional and backward looking it is not perceived as sufficient unto itself. There is a dominant external reality, an alternative viewpoint from which the traditional and parochial is liable to be judged on terms other than its own. There is a "modern world" that may perhaps be joined, or may disrupt and even repress, but that cannot be ignored. Although aspects of premodernity may persist, and indeed be found everywhere, this option is not secure, not a coherent alternative, it can never become hegemonic.

Such a broad generalization necessarily invites counterexamples, and may be clarified by considering a few of the most obvious macro-historical exceptions: arguably the Jesuits in Paraguay before 1763; the Tupac Amaru rebels in the Andes in 1781; the Quilombo de Palmares; the followers of Conselheiro in Canudos in Brazil in 1891; perhaps even the Senderistas in Peru in the 1980s. All of these might be classified as resolutely antimodern bids for hegemony over a substantial slice of the Latin American territory and population. Whether or not one or more of these movements actually invoked elements of an "alternative" view of modernity, what matters here is that all of them were permanently destroyed. From the multiple modernities perspective, the point is that their stigmatization as antimodern (just or otherwise) was decisive in their elimination and in sealing its irreversibility. In a continent constrained by an underlying "bias toward modernity," alternative social proposals and power contenders need the support of elements located in what is generally perceived to be the modern world if they are to maintain a convincing claim on collective aspirations. Only then can they expect to survive and regroup when struck by the periodic setbacks and defeats encountered by any utopian endeavor.

Admittedly, there will be many diverse, overlapping, and competitive strands of external influence and guidance that may be loosely compatible with the aspiration to be modern. Throughout the middle decades of the twentieth century many Latin American opinionmakers came to view Stalin's Moscow as the Mecca of progress. Others opted for the latest encyclicals from the Vatican, others for the most fashionable ideas from Paris. Most recently the World Bank and Harvard have exercised a similar imaginative ascendancy. This range of possibilities is so wide that "multiple modernities" may be thought to embrace everything and exclude nothing. But this criticism would be misplaced on at least two counts: in other large regions of the world we can identify coherent antimodern social constructions that have been or could be more hegemonic than in Latin America; and what Latin America opinion accepts as modern and, therefore, energizing is in fact quite limited and discriminating. At any one time, or in any specific social domain, only a small range of alternative possibilities command the requisite authority. Let us now amplify these two points.

First, we need to establish that other large regions may not all display the same "bias toward modernity." The argument does not require this to be true of *all* other large regions. Indeed, other regions of relatively "recent" settlement display a similar bias (it can be true of North America, Australia and New Zealand, and—a challenging case here—Israel). Nor is the claim categorical. What must be established is a *relative* difference that is sufficiently durable over time and space to distinguish Latin America and the Caribbean from other large world regions. With "globalization" it may be that all large regions are converging, but if so Latin America has been precocious. It has displayed its bias toward modernity for much longer and may, therefore, be less prone to reversion than other large regions that have adopted modernization more recently or less completely.

A classic counterexample is Japan before the forced opening caused by the arrival of Captain Perry's gunboats in 1853. Here was a major civilization that deliberately sealed itself off from the rest of the world and deeply resisted seeing itself through the optic of outsiders. Other examples include imperial China, pre-1895 Korea, Tibet before the arrival of communism, the Zulus before 1879, and Wahabbite Saudi Arabia. All were more inward-looking and resistant to Western modernity far longer than Latin America. These are the most extreme illustrations of a more general syndrome. Ancient civilizations, millennial religions, rich linguistic and cultural traditions, long-standing hereditary elites, all tend to generate their own legitimizing traditions of self-understanding. New external influences that impinge on them tend to be assimilated—or resisted when they cannot be absorbed. By contrast, Latin America's ancient Aztec and Inca civilizations were virtually destroyed, religion was imported from Europe (and latterly from the United States), and even its indigenous movements express themselves mainly in Spanish. The hereditary rulers of Spain, Portugal, and France were all either defeated or rejected in the nineteenth century, but for all that the dominant Latin American pattern of self-understanding has remained outward- rather than inward-looking. Even the population of African origin was cut off from its roots by transportation and slavery, and was subjected to schemas of racial classifications that were constructed and imposed on them from outside their own historical traditions. This is a subcontinent of "peripheral" development, where the question of how one relates to "modern" sources of world leadership occupies center stage in the collective consciousness. Hence the obsessive concern with ratings, rankings, and other evaluations of the countries of the region according to externally constructed criteria and yardsticks. This contrasts with other large regions where the dominant obsession is how the present connects with a majestic historical tradition or a sacred past (Mecca, Ayodhya, Rome, Jerusalem, or indeed Eire or Euskadi).

Second, we need to show that Latin American perceptions of the current locus of modernity in the world are "multiple" but also heavily constrained. Within the "multiple modernities" framework there is one dominant pole of attraction. Paris occupied a pole position throughout the nineteenth century, but the French example exercises no more than a residual influence at the beginning of the twenty-first century; the Soviet image of modernity was even more comprehensively eclipsed. It is the United States that has exerted its fascination as the principal external reference point, first over Mexico and the Caribbean, and later in the twentieth century over South America (although it remains less compelling for Brazil than for most of Spanish America). The rising influence of China may have the potential to expand until it eventually occupies a similar space, but at least for now that is no more than a flickering possibility.

The search for alternatives to this dominant "source" of modernity, and the openness to nonstandard variants of modernization, arises as much from a reluctance to be "Puerto Rican-ised" as from the vigor and power of attraction of rival models. This logic plays out in many arenas. It applies to politics of trade and integration, for instance, with the Summit of the Americas and the FTAA, weakly countered by the Ibero-American Summits, various EU-Latin America common projects, and now by the Bolivarian Alternative for the Americas. It plays out in world cinema, as Hollywood develops its Hispanic menu in competition with Latin American filmmakers whose greatest ambition may be a warm reception in Cannes or an enthusiastic public in Madrid. It can be traced in the choices of undocumented migrants, and the external sources of remittances they send back to their home communities. In almost all areas there is competition, an alternative available to those who do not wish to take the line of least resistance and simply embrace the dominant U.S. model of modernity. In some areas, such as politics or culture, the United States may offer very little choice, but in others (scientific know how, plural sources of information) even the dominant pole of attraction offers the Latin Americans a smorgasbord of possibilities. The flows of foreign direct investment, technical expertise, academic talent, aid, and even email traffic can all be mapped to identify the relative importance of alternative sources of modernity adopted by a wide array of social groups distributed across the whole subcontinent. What such mapping would typically demonstrate is that while the most highly valued external reference points are always multiple, they are also limited in number and normatively constrained. There may be change over time (quite rapid in the case of the suddenly vanished Soviet Union or the forcefully emergent China), configurations may vary in different domains of Latin American social life, and external sources of direction are almost invariably differentiated, but for all that they are few: they are predominantly North American.

Let us now turn to an alternative objection to this way of characterizing Latin America's orientation toward the rest of the world. Reasoning at a very macro and abstract level of generalization, it can be argued that the multiple modernities perspective reduces the rich complexity and innovative capacity of Latin American societies to a question of mere importation or imitation of external models. But I would counter that, on the contrary, this framework helps us understand both the scope and the structure of Latin America's internal debates and experiments. Rich, varied, and creative though they most certainly are, they operate within a specific context of international interactions, feedback, and indeed at times, resistance to lopsided or unduly imitative external models. They are always conducted within a frame of reference that includes Europe and North America (and perhaps other regions as well), and they always pay particular attention to one underlying interrogation: where does our experience fit into prevailing accounts of the structure of modernity? As a consequence of this locally generated but outwardly oriented ferment of ideas, Latin American thinkers and actors typically aspire to make their interpretations count in the modern world as a whole. They will be more convincing, more secure, and more effective at home to the extent that they bend the debate, inject alternative perceptions, and gain a hearing about their own experiences in "advanced" or "developed" or "modern" nations as well.

The multiple modernities perspective not only allows scope for local creativity, adaptability, and resistance to dominant modes of thinking; it also explicitly highlights the potential for *two-way interactions* between Latin America and the main centers of modernity, and the engagement between different variants of modernity as they attempt to mirror Latin America's hybrid social realities. Far from merely imitating an unreflective and uniform modernity established in the advanced Western democracies, Latin American actively contributes to ongoing debates within the modern world, sometimes aligning with one current, sometimes with another, and often reflecting back progressive impulses but with a distinctively Latin flavor. So what is being postulated here is not an unmediated dependency, but a selective appropriation with feedback effects. Thus, the region may back the International Criminal Court or the Kyoto Agreement, or it may embrace the privatization of pension schemes, but it does so according to its own selective perceptions and criteria. And in doing so the fact that it copies external practices is perhaps less relevant than the fact that it also reinforces them, and perhaps even recomposes their constituent elements. The Chilean experience of pension privatization borrowed the ideas of "Chicago School," but it also reshaped them, with consequences for the mature market democracies as well as for its neighboring countries. Similarly, the novels of the "boom" swept Latin America, but they also injected new life into literature on the Iberian Peninsula and throughout Europe. More generally, Latin America's innovative spirit enriches and diversifies modernization tendencies outside the region, as well as drawing inspiration from them. The Latin American presence in multiple arenas makes international modernity more flexible, dynamic, and diverse than it could be otherwise.

Another type of objection to an argument that has been thus far pitched at a very aggregate and general level also requires consideration. A view that is very common among anthropologists, and reflected in this collection by the arguments of Peter Wade, is that sweeping generalizations or master-narratives of this kind carry little substance because they are so far removed from the lived experience of most of the inhabitants of the subcontinent, absorbed as they are in the much more intense and parochial concerns of their local communities. At the small-scale level, where popular culture is mostly experienced and produced, tradition is quite as powerful as innovation, and peer group expectations control behavior more than external reference points critique it. There was a time when many anthropologists privileged and essentialized buried traditions, trying to scrape away the veneer of modern life that concealed what they believed to be deeply rooted and authentic cultural practices of local origin. Current thinking in the discipline no longer favors this approach, but it continues to resist the privileging of modernity as well. Instead, it emphasizes hybridity and the parity of esteem owed to diverse local outlooks, whether primarily adaptive or predominantly resistant to change. For those of Wade's persuasion the anthropologist's rich portrayal of the complexity of social realities at the local level precludes such summative assertions as "Latin America displays an overall bias toward modernity."

From my perspective, however, the arguments sketched out for the region as a whole deserve as much consideration at the micro as the macro level. They cannot be confined to the large-scale picture, but must apply also (if at all) on other scales as well. Guy Thomson's chapter in this volume provides one vivid illustration of how this perspective could be elaborated at a local level, and in a precise historical context. Examples could be multiplied, especially from the "progressive" side of the Latin American cultural divide—the freemasons, liberals, *sanitaristas*, feminists, and military reformers who populate so many narratives of local struggle for emancipation across postindependence history. But the multiple modernities perspective would also invoke the many modernizers of a more conventional stripe—railway builders, coffee planters, monetary reformers, and even the security specialists who also organized and exerted power at local levels in pursuit of their visions of progress and improvement. What these competitive innovators have in common is a sense of the need for change, openness to external models, and ideas about how inherited conditions can and should be upgraded or adapted. This outlook provides the micro-foundations for what I characterize as an overall "bias toward modernity." It is an understandable response to the challenges of Latin America's natural abundance and vast distances; to the destruction of pre-Conquest civilizational traditions; to the overbearing influence first of Europe then of North America; and to the heavily external origins of so much of the credit, technology, and cultural orientations that have driven change throughout the region, even at the most parochial of levels. Of course, such influences are also resisted and adapted at micro as well as at macro level. But over the long run, and in an impressive variety of domains, locally and individually as well as nationally and collectively, Latin Americans have been—and continue to be—strikingly open to images and models of progress and reform validated by the leading societies of the Western world. The practice of judging current circumstances according to an external template, and then agitating for improvement, is deep-rooted and widespread. In this sense Latin America has "always" been modern.

There is limited scope here to illustrate what is being claimed, and demonstration would require discussion of an entirely different order. The claimed "bias to modernity" in the micro-foundations of social practice of the region is a *relative* and *proportional* assertion. It could be defeated by a demonstration that there is as much global reflexivity in Islamic, Buddhist, and Hindu social formations as in Latin America, for instance; that the proclivity to judge and reassess oneself and one's community according to Western criteria of modernity was as great in the former as in the latter. An *absolute* fixation with modernity is, therefore, not required. Similarly, cross-regional demonstrations of this kind would need to weigh up a variety of domains—marriage practices, the sacralization of ancient monuments, responsiveness to commercial fashions, and computer literacy or access to western educational influences. We should expect to find different degrees of reflexivity and unevenness in these various domains across the heterogeneous Latin American social sphere. Even so, the claimed bias would only be disproved if it were shown that, all these variations duly considered, there is no difference in the *overall proportion* of social practices displaying global reflexivity. Different social classes and successive generations would display this characteristic in different ways: with the telegraph, the short wave radio, and the laptop.

It is perhaps not surprising that Latin America's elites of European origin are particularly disposed toward an *extranjerizante* outlook: but my claim goes further, requiring a similar inclination to prevail within the broader popular culture. The keen readership of the *Reader's Digest en Español* in the Mexican village of San José de Gracia in the 1920s and 1930s provides an empirical illustration of the kind of evidence I have in mind. Peter Wade's discussion of the genesis of the music recording industry is another (the fact that this required adaptation as well as importation is not problem from a multiple modernities perspective). Mario Vargas Llosa's reconstruction of the radio shows of the 1950s, or Julio García Espinosa's chapter on the evolution of Cuban cinema fit within the same framework. One of the distinguishing features of Latin America's bias toward modernity (in contrast to the more linear progression typical of the U.S., for example) is that the fitful appropriation of external innovations may lead to "leapfrogging." Thus, for example, the *telenovela* may arrive in an Amazonian village before the school or the post office, emigrant remittances may arrive before potable water, and the Internet may precede a reliable electricity supply. Despite such patchy and uneven development, the bias toward modernity in the subcontinent does extend from the elites to the remotest communities in each nation.

Another central issue here is the "weight of tradition." Sudipta Kavaraj tells me that in Sanskrit there are two clearly distinct concepts of what Westerners elide when they refer to "the past": what has been irretrievably lost, and what survives and is a living source of orientation in the present. In terms of contemporary social practices, the pre-Conquest past is essentially of the first kind and in this it resembles the whole of the "New World" and other areas of "recent settlement." The second kind of "past" is a great deal more prevalent in much of Europe (the "Old World"), Asia, and Middle East and perhaps Africa (African slaves brought across the Atlantic were also deprived of much of their cultural past). Of course, it could turn out to be true, as theorists of modernization and globalization tend to believe, that it will not be long before there is only one present-oriented international society, and that all the distinct outlooks of the various ancient civilizations will become equally remote and irrelevant. In that event, the Latin American bias toward modernity would no longer retain its unique characteristics. But even if the global answer to our "when" question is "now," there are many parts of the world where it can never be said to be "always." In this respect, Latin America is distinctive.

Post-Conquest traditionalism remains a lively presence in Latin American social life, however, and the conquest was five centuries ago. This is an intensely Catholicized culture zone, and surely Catholicism is a deeply traditional worldview and set of archaic practices. So how are we to reconcile this evident "weight of tradition" in the region with the claim that—at least since Independence—Latin America has "always" displayed a bias toward modernity? From a multiple modernities perspective it is an error either to dichotomize tradition and modernity, *or* to spread them out along a smooth continuum. Aspects of the past are selectively appropriated and reinterpreted to suit the purposes of the present. Even the most apparently deep-rooted of backward traditionalisms may, on closer inspection, prove to have been quite recently invented, or even imported. (Consider the *bowler hat*, a distinguishing feature of traditional Andean dress code for women. Think of it!) If a society displays multiple overlapping competing and incomplete projects of modernity, then there must also be multiple, overlapping, and probably incomplete invocations of the past to legitimize them. Hybridity, not unilinearity reigns. The Vatican and the Catholic hierarchy may try to uphold an image of traditionalism and sacrality, but neither Liberation Theology nor the Opus Dei fully conforms to this picture. Both express the hybridity, innovativeness, and openness to external sources of inspiration that I claim characterize this region. The same is true of Protestant Evangelism. Both defenders and opponents of the Cuban Revolution invoke José Martí with equal intensity; Emiliano Zapata's legacy is still contested between PRI-istas and repackaged Marxists; and in Ecuador *¡Alfaro Vive, Carajo!*. Beyond the strictly political sphere tangos, *corridos*, gaucho poets, *niños héroes*, and the founders of legal doctrines are all claimed, appropriated, and distorted for current use. Multiple pasts are mobilized to orient ongoing debates between alternative futures.

This volume ranges across a multiplicity of domains that are normally studied in isolation from each other—politics, cinema, local history, and cultural practices. My characterization of Latin America is intended to sum up distinctive features of all these domains and many others—the economy, international relations, architecture, and collective memory. Such a wide focus invites the inevitable criticism that generalizations derived from knowledge of one or a few of these domains is inapplicable to the rest. It must certainly be conceded that the balance between tradition and modernity (if it can be ascertained at all) is likely to vary between architectural practices and *telenovelas*, or music and politics. But if, as the multiple modernities perspective and this volume presume, there is strong interaction and indeed permeability between these domains in a fluid and modern-oriented society, then it is still possible to defend our summative, holistic, generalization.

Permeability between domains certainly operates in both directions. If the Internet helps Latin America to "modernize" in the sense of conquering distance (a long-standing major barrier) and speeding up information flows, this not only helps Zapatistas to overcome the isolation of Chiapas and to mobilize intellectual opinion in the most fashionable parts of Europe against the "neoliberal" project of the Mexican state (thus counterpoising one project of modernity against another), the same instrument can also help the Vatican tighten its grip on the region's church hierarchy and curb the freedom of action of progressive clerics. Such sector-specific exchanges are taken up by the broader institutions of modern society. They reappear in *telenovelas*, in film, in the discourse of public intellectuals, and even in university courses. One of the striking features of the subcontinent's orientation toward modernity is the fluidity of communication between diverse social arenas. Latin American societies are characteristically open to heterogeneous and cross-cutting ideas and fashions, both from within and without. This provides the social foundation for recurrent bursts of what is often described as "populism"—a somewhat incoherent combination of hopes and aspirations from diverse sources brought together (at least temporarily) by a unifying leader, or by a "movement" that unites *against* a nonparticipating target group, rather than in favor of a solid programmatic interest. Beyond the political realm, it is apparent in other domains such as the socially and even racially unifying symbolism of the national football team, the trans-class appeal of Latin American popular music, the cross-national stature of famous writers and thinkers, the receptivity to visitors and immigrants from the metropolitan centers of modernity, and the support networks linking overseas migrants to their families and communities of origin. This openness, mobility, and cosmopolitanism reflect the interplay of the multiple modernities characteristic of the subcontinent, and can be contrasted to the more closed, defensive, and inward-looking social orientations found in various other large global regions, where the domains of religion, culture, politics, economics, migration may fit together more tightly, cumulatively repelling cosmopolitan tendencies and entrenching self-contained social practice and traditions. In Latin America, broadly speaking, such domains are more loosely integrated; they operate with more hybridity and permeability (think of the *sambadrome*, of *capoeira*, of the *peña folclórica*); there are variations of scale that invite experimentation outside one's closed social circle (privileged intellectuals become peasant insurgents, school dropouts become *piqueteros* or even *maras*, and indigenous women win Nobel Prizes). All this leaves a relatively open arena for variation, innovation, and adaptation within the general framework provided by peripheral modernity.

So, "When was Latin America modern?" Some questions invite categorical answers ("When did Columbus first make landfall in the Americas?"); others can only be tackled perspectivally ("How blue is the Caribbean?"). Both types of question deserve scholarly replies if they address important issues about our understanding of the world we inherit. "When was Latin America modern?" is a question worth investigating, even though it belongs in the second of the two categories. Like the luminescence of the Caribbean, or the pursuit of happiness, it addresses a palpably significant issue but not one that is self-evidently answerable through conventional metrics. The Caribbean is amazingly blue. Those familiar with the North Sea or the South Atlantic can hardly fail to acknowledge the difference. Yet this aquamarine radiance is not uniform. There are hurricanes, and even a few wintry nights. Its blueness is not absolute or unvarying over time. Some observers find it impossible to miss whereas others pay it little heed. So, like the precocious modernity of Latin America, it is a relative and perspectival truth, but no less "real" for all that.

It might be thought that to answer the question "When was Latin America modern?" with "always, since the region has a recurrent bias toward modernity" involves a sleight of hand. The question seems to invite a precise definition leading to a categorical answer that gives dates and uses an objective metric, whereas the response could be dismissed as unspecific and too interpretative, even "structuralist." Such objections merit consideration, although my position is that there is no sleight of hand, just a more illuminating way of thinking about the issues underlying the question. The response is broader than the question, and does lack the precise timing, the fixed boundaries, and the unambiguous measurement criteria implied by its wording. But this type of response can still be defended against its critics, of which two groups can be discerned elsewhere in this volume. The first group veers toward logical positivism, in that they are incline to reject as either meaningless or confused any response that eludes their strict tests for reliability and significance. The second group is so relativist and/or subjectivist that they reject the epistemological assumptions they attribute not only to the question but to all direct attempts to answer it. Both types of criticism blunt our capacity to understand the social world around us.

The first group demands precise definitions and objective measurements, and their search for precision can yield some helpful initial results. But while precision is invaluable when examining, for example, such questions as what is specifically "modern" about the "modern state," too narrow an emphasis on exactitude would foreclose exploration of the broader issues posed by "modernity" in general. Consider the parallel questions of method posed when we evaluate claims about "happiness." "I'm happy when my team wins," is an empirically checkable statement with a reasonably precise meaning. "The pursuit of happiness is a basic human motivation," is another highly meaningful claim, but it operates at a much higher level of abstraction and is correspondingly harder to verify, falsify, or even specify. But the human sciences would be impoverished if they could only address the first of these two claims and were precluded from examining the larger assertion. Now let us switch back to modernity. In the same way that the "pursuit of happiness" can be understood as a recurrent subjective aspiration that can never be definitively fulfilled or even unambiguously measured, so also a "multiple modernities" perspective will resist categorical definitions and final closures. That is why neither can be dismissed as teleological. No such objection can apply when "modernity" refers to a collective aspiration, a succession of overlapping and competing objectives for improvement, a restless urge for experimentation that can never be satisfied. Such a conception of modernity is neither meaningless nor muddled. In fact it should even be possible to assemble objective indicators that track its presence or absence. But this conception is both relational and perspectival, and as such it will tend to elude the criteria of precise objectification required by the first set of critics. It is *relational*, because the notion of tradition is the essential counterpart to that of modernity (while bearing in mind that all traditions were once modern inventions, just as all durable modern practices are destined in time to be seen as traditional). It is *perspectival*, because any aspiration (or bias) toward modernity must be grounded on a view of where we are now and what needs to be changed—intersubjective views that are never simply given, but that have to be constructed and are always open to renegotiation.

The second (relativist/subjectivist) group takes up these valid points about the relational and perspectival features of modernity, but overdoes them. Those of this viewpoint seem to think it wrong in principle to distinguish this region's "orientation toward modernity" from that of any other world region. They object to what they regard as the implied assumptions of hierarchy, unilinearity, and teleology. They oppose privileging the viewpoint of one type of observer (the macro classifier) to the detriment of all other standpoints. Against these critics, a "multiple modernities" perspective would insist on the merit of identifying certain sites of modernity (for example, the Internet) in contrast to other sites of tradition (say, Talmudic scrolls). Once this first step is taken then it becomes logical to differentiate between those social contexts (including those large regions) where sites and practices are more densely concentrated and exercise more grip on the collective imagination; and those where the reverse is the case. If so, then it becomes appropriate, at least in principle, to attribute a "recurrent bias toward modernity" to regions where over long periods of time we keep finding the first pattern rather than the second. This exercise in classification can help us to explain the social world around us, and need not necessarily involve imposing some artificial and one-sided worldview.

There is more than one way to establish the veracity of my main thesis. Various types of measurement can be considered—the proportion of the population using the Internet, or receiving remittances, for example (or the hours of sunshine to which an ocean is exposed). But from a perspectival standpoint the frequency of Caribbean cruises, or the price of waterfront real estate, may provide more solid evidence of what is (socially) "real." Intersubjectively shared beliefs and perceptions can produce self-validating behavioral consequences. Latin American self-constructions as belonging to the Western world, as wide open to modern innovations, and as subject to external evaluations, all make a difference to collective outcomes, and all differentiate this large region from various others. Such self-constructions and their consequences become integral to the region's sense of identity, and become part of the "tacit knowledge" used by local actors and their external interlocutors alike in their routine praxis. For this reason, the question "When was Latin America modern?" is not only meaningful, but also urgent and relevant to many contemporary debates. Although not usually articulated so explicitly, it lurks behind many anxious discussions about the region and its prospects, debates that are constantly renewed both within and outside it. Confronting it head on, as Nicola Miller's and Stephen Hart's initiative has pressed all of us to do, encourages us to pull together normally disconnected areas of scholarly enquiry and reflection. It also obliges us to spell out our underlying methods, assumptions, and perspectival viewpoints. Mine is that a multiple modernities perspective supports a holistic conclusion. Just as the Caribbean is very blue, so Latin America has long been precociously, unevenly, incompletely, but for all that, decisively modern.

Notes

1. See Sankar Muthi's careful recent reading of the main Enlightenment authors, which reveals

    a genuine and contentious struggle among eighteenth-century thinkers about how to conceptualize humanity, cultural difference, and the political relationships among European and non-European peoples. Indeed, on many topics, a wide range of eighteenth-century thinkers, many of them well known and influential, posit theories that fail to fit the "project" or the core intellectual dispositions attributed to modernity in general or to "the Enlightenment" in particular. . . . It may well be the case that only a negative definition of Enlightenment thought, one based on what "Enlightened thought is against," could underlie the extraordinary plurality of texts, arguments and dispositions that one can find in such Enlightenments . . . . (Sankar Muthi, *Enlightenment Against Empire* Princeton University Press, Princeton, 2003: 264–5)

2. Of course "Latin America" did not exist as a category or region before the 1860s. It was a "modern" invention of the French Second Empire. "Indo-America" is an even more modern twentieth-century contrivance. The inclusion of the Caribbean is historically and geographically necessary but complicates the picture.

3. Laurence Whitehead (2006), *Latin America: A New Interpretation*. New York: Palgrave.

List of Contributors

João Cezar de Castro Rocha is Professor of Trans-Atlantic Comparative Studies at the University of Manchester, UK. He was previously Professor of Comparative Literature at the Universidade do Estado de Rio de Janeiro and has published widely on Brazilian Literature and Culture in comparative context. His book *The Author as Plagiarist: The Case of Machado de Assis* was published in 2006 by Luso-Brazilian Books.

Néstor García Canclini is Research Professor at the Universidad Autónoma Metropolitana, Mexico City. Renowned for the classic work of Latin American Cultural Studies, *Hybrid Cultures: Strategies for Entering and Leaving Modernity*, which has recently been republished with a new introduction (University of Minnesota Press, Minneapolis and London, 2005). His other works include *Consumers and Citizens: Globalization and Multicultural Conflicts* (University of Minnesota Press, Minneapolis and London, 2001).

Julio García Espinosa is Director of the Escuela Internacional de Cine y Televisión, San Antonio de los Baños, Cuba. Internationally renowned as a director and screenwriter, some of the most acclaimed films from his long and distinguished career include *El joven rebelde* (1962), *Las aventuras de Juan Quin Quin* (1967), *De cierta manera* (1977), *Reina y rey* (1994), and *Enredando sombras* (1998).

Stephen Hart was educated at Downing College, Cambridge and is Professor of Hispanic Studies at University College London where he teaches courses on Latin American Literature and Film. He has published *A Companion to Spanish American Literature* (1999), *A Companion to Latin American Film* (2004), and coedited Contemporary Latin American Cultural Studies (2003). He holds an honorary doctorate from the Universidad Nacional Mayor de San Marcos and has been awarded the Orden al Merito by the Peruvian government for his work on Cesar Vallejo.

Alan Knight is Professor of the History of Latin American and Fellow of St. Antony's College at the University of Oxford, UK. He has published extensively, especially on the history of Mexico, including *The Mexican Revolution* (Cambridge University Press, Cambridge and New York, 2 vols., 1986); *US-Mexican Relations, 1910–1940* (University of California Press, San Diego, 1987) and two volumes of a three-volume general history of Mexico: *Mexico: From the Beginning to the Conquest*, and *Mexico: The Colonial Era* (Cambridge, 2002).

Nicola Miller is Reader in Latin American History at University College London. Her previous books include *Soviet Relations with Latin America, 1959–1987* (CUP, 1989) and *In the Shadow of the State: Intellectuals and the Quest for National Identity in Twentieth-Century Spanish America* (Verso, 1999). She is currently working on a history of intellectuals and modernity in Latin America.

Sarah A. Radcliffe is Senior Lecturer in Latin American Geography and Fellow of New Hall at the University of Cambridge, UK. Her publications include *Culture and Development in a Globalizing World: Geographies, Actors and Paradigms*, ed. (Routledge, London, 2006); *Re-Making the Nation: Place, Politics and Identity in Latin America*, with Sallie Westwood (Routledge, London, 1996); and *Viva: Women and Popular Protest in Latin America*, ed. with Sallie Westwood (Routledge, London, 1993).

William Rowe is Anniversary Professor of Poetics in the Department of Spanish and the School of English and Humanities at Birkbeck College in the University of London. His many publications on Latin American Literature and Culture include *Memory and Modernity: Popular Culture in Latin America*, with Vivien Schelling (Verso, London, 1991); and *Poets of Contemporary Latin America: History and the Inner Life* (Oxford University Press, Oxford, 2000).

Guy Thomson is Reader in History at the School of Comparative American Studies, University of Warwick, UK. His books include *Puebla de los Angeles. Industry and Society in a Mexican City, 1700–1850* (Westview Press, 1989); *Politics, Patriotism and Popular Liberalism in Mexico Juan Francisco Lucas and the Puebla Sierra 1854–1917* (SR Books, 1999); and *The European Revolutions of 1848 in the Americas* (Institute of Latin American Studies, London, 2001).

Peter Wade is Professor of Social Anthropology at the University of Manchester, UK. Recent works include *Race and Ethnicity in Latin America* (Pluto Press, London, 1997); *Music, Race and Nation: Música Tropical in Colombia* (University of Chicago Press, Chicago, 2000); and *Race, Nature and Culture: An Anthropological Approach* (Pluto Press, London, 2002).

Laurence Whitehead is Official Fellow in Politics at Nuffield College, Oxford University, UK. He also became the first Director of Oxford University's new Centre for Mexican Studies in 2002. Among his most recent publications are *Latin America: A New Interpretation* (Palgrave, New York, 2006); *Democratization: Theory and Experience*, ed. (Oxford University Press, Oxford, 2002); and *Emerging Market Democracies: East Asia/Latin America*, ed. (Johns Hopkins University Press, Baltimore 2002).

Index

Adán, Martín (De la Fuente
Benavides, Rafael), 127,
128–31, 143 (n. 5)
Afghanistan, 26–7
Africa, 30, 56, 58, 61–2, 72, 73, 86
(n. 6), 200
cinema in, 169–70
Alvarez, Santiago, 169, 175 (n. 11)
Amado, Jorge, 154
Amazon
cartography of, 36, 191
indigenous peoples, 58–9
modernity in, 200
American Independent Cinema, 174
Anderson, Benedict, 36
Anderson, Lindsay, 174 (n. 9)
Añez, Jorge, 63
anti-modern movements, examples
of, 193
Argentina, 7, 30, 63, 104,
114 (n. 35), 171, 178, 180
Arguedas, José María, Los ríos
profundos, 134–8, 143 (n. 12)
Armento, Roberto, 77
Asia, 71, 73, 200
asymmetry of power, between
North and South, 11, 13,
184, 188
Australia, 14, 185, 194
authorship, 9, 152, 154–8
avant-garde movements, 14

Barber, Benjamin R., on concept of
totalitarianism, 96
baroque modernity, critique of,
105–6

Barth, John, on Machado de Assis,
159
Barthes, Roland, 3, 126
Basadre, Jorge, 124–7, 133,
142 (n. 1)
Batista, Fulgencio, 10, 168, 174 (n. 7)
Baudelaire, Charles, 105,
114 (n. 39), 132
Bauman, Zygmunt, 23, 32
Bayly, C. A., The Birth of the Modern
World 1780–1914, 6, 71–4, 85
definition of the modern, 71
discussion of Hispanic world,
6, 72–4
Belaúnde Terry, Fernando, 122
Bell, Daniel, 50
Bell, Morag, 44 (n. 21)
Benjamin, Walter, 8, 122, 130, 132,
139, 141, 143 (n. 6),
144 (n. 19), 171
Berman, Marshall, 50
Beverley, John, 55, 56
Bloom, Harold, on Machado de
Assis, 152, 156
Bolívar, Simón, 134
Bolivia, 23, 31, 144 (n. 16)
Borges, Jorge Luis, 148, 152, 154,
155, 160
Bourbon Reforms, 108
Bourbon Restoration, 75, 85,
86 (n. 4)
Brazil, 9, 34, 37, 92, 148–9, 171,
195
ideas about race, 60, 61
indigenous peoples of, 59
see also Candomblé

Briceño, Alcides, 63
Brodsky, Joseph, 147

Caldwell, Helen, 156
Canada, 178, 194
Canary Islands, musicians from,
    63–4
Candomblé, 61–2
capitalism, 2, 5, 12, 24–5, 41, 50,
    55, 57, 58, 59, 99, 101, 102,
    107
"Hispanic", 73
cartography, 26–7, 33, 35, 36,
    181–2, 187–8, 191
Castro Rocha, João Cezar de,
    chapter summary, 8–9
Catechism of Geography of the
    Republic of Ecuador (Mera), 35
Catholic Church, 76–7, 80, 84, 115
    (n. 43)
Chile, 65, 104, 114 (n. 35)
    pension privatization, 197
China, 14, 72, 76,102, 194, 195
cinema
    in Cuba, 9, 167–76
    in Latin America, 184–7, 196
    in USA, 184–7
Cliometrics, 111
Cold War, 10, 37, 168
Colombia, 32, 37
    music in, 5–6, 59, 63–4
colonialism, 28–9
    consequences in Latin America,
        31, 72
Columbia Gramophone Company,
    62–3
Communist International, 141
Conrad, Joseph, The Nigger of the
    Narcissus, 132–3
Cornejo Polar, Antonio, 136–7
Coronil, Fernando, 25, 28, 40
Corsín Jiménez, Alberto, 65
Cuba, 7, 200
    cinema in, 9, 167–76
    Cuban Film Institute (ICAIC), 9,
        170

Havana Film Festival, 170
International Film and TV
    School, 170
cultures of consumption, 6, 7, 55,
    76, 78–9, 84
Cunha, Euclides da, Os sertões,
    149–50

Dalby, Simon, 26–7
De la Fuente Benavides, Rafael (aka
    Adán, Martín), 127, 128–31
democracy, 2, 99, 107
democratic optimism, 75–8
development, 23, 25–6, 37, 38–40
Díaz Quesada, Ernesto, 167
Díaz, Porfirio, 74, 84
discontinuity, in Latin American
    history, 12, 29
dualisms, 5–6, 12, 29, 51, 59–60,
    65
Dunkerley, James, 51, 60, 70
Dussel, Enrique, 29, 52

Eakin, Marshall, 149
Ecuador, 27, 30, 31, 34–8, 201
    cartography, 36
    Cochasqui (archaeological site),
        38
    currency, 35
    Escuela Nacional Politécnica, 35
    Geographical Military Service;
        later, Geographical Military
        Institute (IGM), 36
    military, 36–7
    national anthem, 35
    National Cultural Council, 37
    railways, 35
    regionalism, 38
    roads, 38
    war with Peru (1941), 37
Eguren, José María, 127–8
El Espíritu Público (newspaper,
    Spain), 78
emic/etic distinction, 7, 8–9, 93
Englund, Harri and Leach, James,
    56–7

Enlightenment, the, 2, 7–8, 10, 23, 24, 29, 30, 42, 50, 100–4, 105, 108, 109, 192, 205–6 (n. 1)
Enzensberger, Hans Magnus, 179
Escobar, Arturo, 39, 40, 50, 52
Eurocentrism, 4, 5, 8, 13, 21, 25, 38, 51, 52
Europe, 10, 13, 21, 30, 73, 75, 76, 96, 100, 102, 178, 200

Fabian, Johannes, *Time and the Other*, 57
favela
    etymology, 150
Flores Galindo, Alberto, 124
    La agonía de Mariátegui, 138–41
Forment, Carlos, 7, 113 (n. 27)
Foucault, Michel, 22, 30, 33, 105
Free Cinema, 174
French Revolution, 50, 139
Fuentes, Carlos, 153, 154, 162 (n. 18)

Gaonkar, Dilip, 131–2
García Canclini, Néstor, 13, 30, 50
    chapter summary, 10–11
    Culturas híbridas/Hybrid Cultures, 5, 52, 55–6
    Latinoamericanos buscando lugar en este siglo, cover design, 187
García Espinosa, Julio, 200
    chapter summary, 9–10
García Márquez, Gabriel, 170
García Moreno, Gabriel, 35
Garcilaso de la Vega, El Inca, 134–5
Garibaldi, Giuseppe, 77, 80
Geertz, Clifford, 113
gender, 30–1
    in Latin America, 24
Geografía de Ecuador (school textbook, Terán), 37
Giddens, Anthony, 24
Gilroy, Paul, 61
Gledson, John, 158

globalization, 1, 54, 55, 181, 182, 194, 200
    archaic, 72
González Prada, Manuel, 122, 131
Gregory, Derek, 28, 30
Gutiérrez Alea, Tomás, 9, 10, 168

Habermas, Jürgen, 23, 49
Haiti, 92, 110, 191
Hardt, Michael, and Antonio Negri, *Empire*, 32, 49–50
Harvey, David, 12
Haya de la Torre, Víctor Raúl, 124
heterogeneity, 137, 140, 192
Hollywood, 172, 185–7
Hoosen, David, 36
human agency, 5, 25, 30, 54
Humboldt, Alexander von, 108
hunter-gatherers, debates about, 57–9
Huntington, Samuel, 92
hybridity, 5, 6, 51, 55–6, 59, 82, 192, 197, 198, 201, 202

Independence wars, 2, 12, 108–9
indigenous peoples, 31, 36, 43 (n. 12), 58–9, 72, 79, 81–2, 84, 94, 195
Industrial Revolution, 50, 130
industrialization, 93, 102
Instituto Pan-Americano de Geografía e Historia, 37
intellectuals, 11, 40, 42 (n. 3), 60, 61, 62, 95, 125, 140, 172, 202
Inter-American Geodesic Service, 37
interdisciplinarity, 3–4, 13–14
International Working Mens' Association, 85
Internet, 201
Israel, 194
Israel, Jonathan, 112, 113
Italian neo-Realism, 10, 168
Italian Risorgimento, 80–1, 84
Italian Unification, 76

Japan, 14, 194
Joseph, Gilbert, 34
Joyce, James, *Ulysses*, 159
Juárez, Benito, 77, 80

Kalahari desert, 58
Kavaraj, Sudipta, 200
Keynes, John Maynard, 100, 101, 102, 112 (n. 24)
King, Anthony, 52
Knight, Alan, 8, 11, 15
  chapter summary, 7
Koselleck, Reinhart, 15
Kossuth, Lajos, 77
Kuper, Adam, *The Invention of Primitive Society*, 57

Landes, David, *The Wealth and Poverty of Nations*, 73
Larraín, Jorge, 40
Lash, Scott, 50
Latin America
  definition, 92
  stereotypes, 73–4
Latin American Film Directors' Committee, 170
Latin American Film Festival, 175
Latour, Bruno, *We Have Never Been Modern*, 14, 51
Lefebvre, Henri, 27
Leguía, Augusto, 122, 123
Lewis, Oscar, 86
Lida, Clara, 74
Lima, modernization of, 123
Lincoln, Abraham, 77, 80
literary "boom", 197
Lomnitz, Claudio, 12
Loveman, Brian, 73
Lucas, Juan Francisco, 82
Lyotard, Jean-François

Mac Adam, Alfred, 152
Machado de Assis, Joaquim Maria, 9, 149, 151–60
  Dom Casmurro, 155–6

Posthumous Memoirs of Brás Cubas, 151–2
Quincas Borba, 154–5
Ressurreição, 152
maps, *see* cartography
Mariátegui, José Carlos, 8, 122, 123, 124, 138–41
Martí, José, 173 (n. 1), 201
Martín-Barbero, Jesús, 132
Marxism, 38, 102, 124
Matory, J. Lorand, 61–2
Mazzini, Giuseppe, 77
McIntyre, Alasdair, 112
Melanesia, 56–7
Mera, Juan León, 35
mestizaje [racial and/or cultural mixing], 5, 56
Mexico, 7, 27, 30, 70–1, 74–84, 92–3, 195, 199
  border with USA, 182–3
  industrialization, 92–3
  National Guard, 82, 84
  Puebla Sierra, 6, 74, 78–9
  railways, 78
  remittances from USA, 181
  "socialist" education, 102
  US anthropologists in, 6, 70
Middle East, 28
Mignolo, Walter, 29, 33, 52, 54
migration, 10, 11, 61–2, 178–81, 196
modern versus non-modern states, 2
modernism, 14, 50, 69, 105, 177
modernismo, 2, 69
modernity
  alternative forms of, 51, 131–2;
    see also modernity, Latin American
  anthropological approaches to, 5–6, 50–2, 60
  as discourse, 23, 28–30, 32
  as experience, 24, 32
  as project, 22–3, 24–7, 32
  aspirational aspects of, 13–14
  associational life of, 6, 80–1
  chronology of, 2, 21, 49–50, 111
  criteria of, 98–9

cultural studies approaches to,
13–14, 105
emancipatory promise of, 10, 11,
12
embodiment of, 5, 30–1
etymology of, 49
geographical approaches to, 4–5,
22–34
geographical practices of, 26–7,
32–3,
historical approaches to, 6–7, 8,
13, 69
Islamic, 41, 43
Latin American; agents of, 198;
alternative forms of, 3, 9, 10,
52, 142; chronology of, 2–3,
97, 191; scholarship on, 1–2;
use of term, 7, 69, 94–5
literary approaches to, 13–14, 86
social science approaches to, 14,
104–5
validity as analytical category, 1, 4,
5–6, 7–8, 41, 51–3, 65, 85,
91, 106–10, 193–5, 202–5
see also multiple modernities *and*
scaling *and* teleology
modernization theory, 38–9, 98,
104, 106–7, 192
modernization, 69, 177
Moreiras, Alberto, 50, 54
multiculturalism, 11, 188
multiple modernities, 1, 4, 12–13,
22, 25, 26, 32–3, 40, 41, 51,
56–7, 64–5, 192–205
music industry, in Latin America,
62–4
music, in Colombia, 5–6, 59, 63–4
Muthi, Sankar, 205

Narváez, Ramón María, 81
National Security Doctrine, 37
nation-states, 5, 12, 24, 26, 34–5
eclipse of, 10
natural selection, 5, 53–4
neoliberal modernity, 10, 23, 104
neoliberalism, 2, 26, 104

New Latin American Cinema
Foundation, 170
New Latin American Cinema, 169
New Wave (Nouvelle Vague), 174
New Zealand, 14, 194
Nigeria, 62
Norman Conquest, 103
Nugent, Daniel, 34

Occidentalism, 28
Odría, Manuel, 122
Ogborn, Miles, 27, 29
Oporto, Henry, 23
Orientalism, 28
originality, 9, 153–4, 160

Padrón, Juan, 169
Palma, Ricardo, *Tradiciones
peruanas*, 122–3
Paris Commune, 85
Paris, 195
Park, Robert, 71
Parsons, Elsie Clews, 70
Peón, Ramón, 168
Peru, 7, 8, 31, 121–45
provinces versus Lima, 125–6
Poole, Deborah, 60–1
populism, 95–6, 202
positivism, 35, 122, 159
postcolonial studies, 12, 28–30, 38,
44 (n. 21), 52, 74
post-development theory, 39–40
postmodernism, 1, 105–6, 107
postmodernity, 14, 105, 193
PRATEC (Peruvian NGO), 39
progress, 8, 34, 75–6
Protestantism, 80, 82, 83, 101–2,
201
Puerto Rico, 195
Pulido, Juan, 64

Quijano, Aníbal, 52, 134–6

race, 30–1, 60
Radcliffe, Sarah, 11, 12, 13, 15
chapter summary, 4–5

Ramírez Erre, Marcos, 183–4
rationalism, 100
rationality, 2
  alternative, 136
Redfield, Robert, 70–1
Reformation, the, 2, 50
Reiss, Wilhelm, 35
Renegger, N. J., 112–13
Revolutions of 1848, influence on
  Hispanic world, 77
Ribeyro, Julio Ramón, 123
Riva Agüero, José de la, 123
Rosaldo, Renato, 2, 55, 58
Rostow, W. W., Stages of Growth, 39
Rowe, William, 11, 13
  chapter summary, 8

Sá Rego, Enylton de, 158
Said, Edward, 28, 54
Salazar Bondy, Sebastián, Lima la
  horrible, 123
San people (aka "Bushmen"), 58
Santa Anna, Antonio López de, 86
Santiago, Silviano, 155
scaling, 1, 5, 6, 7, 9, 13, 31, 33,
  51–7, 60, 64
Sarmiento, Domingo Faustino, 94,
  109
  as read by da Cunha, 150
Schwarz, Roberto, 148, 158
scientific revolution, 2
secularism, 101, 103
Sekula, Allan, 186–7
Seville Exhibition (1929), 63
Shakespeare, in work of Machado de
  Assis, 156–8
Siegel, Micol, 61
Smart, Barry, 49
Smith, Neil, 25
sociability, new forms of, 80–1
Socialist Realism, 175
solidarity, 6, 10, 11, 81–4,
  188
Sontag, Susan, 159–60
sovereignty, 26–7, 142
Soviet Union, 14, 195

Spain, 6, 29, 43, 75–85
  carbonari societies, 82–3
  Democrat Party, 76, 77
  labour practices, 83
  Málaga-Granada highlands, 6,
    74–5, 78–9
  National Militia, 82
Spanish Conquest, 103
spatiality, 8, 11, 21, 24, 26, 27, 32–3,
  36, 38–41, 42 (n. 5), 50, 52–3
Stendhal, 151, 152
Sterne, Laurence, 156, 159
Stiles, Daniel, 58
stretched-out geographies, 30, 31,
  33, 41

Taussig, Michael, 57, 58–9
Taylor, Charles, 14, 50
technology, 2, 75–6, 101, 109,
  124, 136
teleology, 1, 4, 5, 6, 7, 9, 13,
  51–7, 60, 64, 69, 92,
  192, 204
temporality, 8, 11, 50, 52–3,
  121–44, 150–1, 192
Terán, Francisco, 36, 37
Thomson, Guy, 93, 198
  chapter summary, 6–7
Thurner, Mark, 3
Titanic, filming of, 186–7
Torres García, Joaquín, map of
  South America, 181–2
tradition, 5, 9, 39, 55–6, 59, 71, 95,
  100, 114 (n. 36), 193, 195,
  198, 200–1, 204
transculturation, 137
translation, 136–7
Turner, Bryan, 49

uchrony, 124, 126, 139, 140
United States, 102, 110, 194
  as model for Latin American
    modernity, 10, 181, 196
  as modernized nation, 2, 103, 200
  beliefs in, 113–14
  border with Mexico, 182–3

film industry, 185–6
multiculturalism in, 11, 184
race relations, 60, 61
untranslateability, 136–7
Uruguay, 104

Valdelomar, Abraham, 125, 134
Vallejo, César, 139, *España, aparta de mí esta cáliz*, 141–2
Vargas Llosa, Mario, 132, 199–200
  La casa verde, 132
  Conversación en la Catedral, 132–4
  Pantaleón y las visitadoras, 132
Venezuela, 171
Vich, Víctor, 137

Victor Talking Machine Company, 62–3
Viola, Bill, 129

Wade, Peter, 1, 11, 107, 198, 199
  chapter summary, 5–6
Wagley, Charles, 148
Whitehead, Laurence, 13, 97, 101
Wilkie, James, poverty index, 94
Wolf, Theodore, 35

Yanagi, Yukinori, 182, 183–4
Yoruba culture, 62

Zapata, Emiliano, 201
Zapatistas, 32, 201
Zavattini, Cesare, 168

Printed in the United States
By Bookmasters